Tackling Biology Projects

Marilyn Anne Wedgwood

BSc, MSc, PhD, CBiol, MIBiol

MACMILLAN

First published 1987
Reprinted 1991

Published by
MACMILLAN EDUCATION LTD
Houndmills, Basingstoke, Hampshire RG21 2XS
and London
Companies and representatives
throughout the world

Printed in Hong Kong

British Library Cataloguing in Publication Data
Wedgwood, Marilyn
Tackling biology projects. – (Tackling
projects)
1. Biology
I. Title   II. Series
574   QH308.7
ISBN 0–333–42869–2

Library

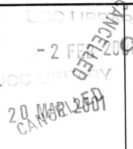

*Dedicated to the memory
of my father
Alexander Mackay Craig*

Contents

Preface

I wrote this book in the hope that it would give direct help to students who have to complete a practical project in Biology – thus fulfilling an often expressed desire by students for more guidance. The idea evolved as I witnessed the difficulties and problems many students faced when embarking on a project, and as I recognised that such problems and difficulties were similar between different groups of students and recurred each new academic year. Furthermore it became clear to me from the comments made by supervisors and examiners that there existed many common mistakes by students in the approach to the project work across a range of different types of projects, and students too often commented that they would have appreciated some clear, straightforward, practical guidance to enhance the experience gained.

This book therefore aims to give straightforward, easily accessible and relevant help and guidance on how to approach and do a Biology practical project successfully. It is written primarily for undergraduate students on BSc, GIBiol (MIBiol), B/TEC HND and HNC courses, though postgraduates embarking on an MSc or PhD might also find it useful.

As I wrote the book I tried to ensure that the advice offered was consistent with the generally established view of sound scientific practice, took account of the common problems and mistakes made by students and kept sight of the fact that the project is an assessed part of a course. Obviously, in a book of this size and approach it is impossible to give a full examination of many of the issues but there are other more detailed and specialised texts which can be referred to. I would, however, be most happy to receive comments on the usefulness of the book and any suggestions for modifications and improvements.

A bibliography has not been prepared as it was felt to be generally inappropriate within the context of the aims of this book. There are many related areas which could be pursued through the subject index of a library under such headings as: experimental design; experimental methods; scientific methods; writing reports; practical investigation; projects in Biology; and philosophy of science.

I am grateful to the many students at Manchester Polytechnic who acted as 'guinea pigs' for testing out some of the ideas – often unwittingly, and for their comments and reactions to the work – particularly Carl Brown and Craig MacKerness for their constructive criticism. I would particularly like to thank Mantz Yorke for his encouragement and patience in ploughing carefully through the early manuscripts and Vic Pentreath for his valuable criticism and advice which concentrated and focussed my thoughts. I am grateful to Roger Williams for reading, advising on, and correcting the chapter on statistics, to Michael Head, John Gaffney and Derek Gordon for comments on the early chapters, and to Jean Higginbottom and Tina Theis for their advice on chapter 7. I would like to convey my special thanks to my friend and colleague, Maureen Dawson, who offered support and advice throughout with good humour and good sense, and who generously read and made useful comments on the final manuscript, and to my husband Barry for his continual encouragement and reassurance which sustained me through the various crises of confidence, so enabling me to complete the work. I would like to thank Christine Foster who typed the final manuscript caringly and carefully, and Marion Sibbit for patiently typing some of the earlier drafts – each of whom worked from some untidy and almost illegible scripts. Last but not least, I would like to thank Gary Larvin for the good humoured drawings.

How to Use this Book

The flow diagram which follows gives you a summarised step by step guide to completing your Biology project. Use the flow diagram as an overall guide and main point of reference as you tackle your project.

Four stages to the project are identified in the flow diagram: Initial Preparation; Design the Investigation; Do the Practical Work; and Write the Project Report. For each stage there is a series of tasks which you will have to complete. References to sections in the book are given for each of these tasks, to help you. The outcomes resulting from completing the tasks are identified in the flow diagram.

Each chapter of the book is divided into sections and the sections have two main styles of text. There are checklists of direct practical advice for easy and quick reference. Other text in the section then discusses some of the issues in more detail.

At the beginning of the book (sections 1.1, 1.2, 2.1, 2.2 and 2.3) some more general aspects of the project are discussed. They are worthwhile reading for gaining a fuller understanding of the project and for using other parts of the book. However, they are short on the direct practical advice that the other sections concentrate on.

The book can be approached and used in various ways. The chapters are presented in an order which reflects as far as possible the way in which a project develops. You could, therefore, follow it through chapter by chapter with the flow diagram guiding you. However, it is possible to delve into the different chapters and sections as the need or your interest arises. (The sections are as self-contained as possible, with cross-referencing, and a full index is given at the end.) Alternatively, you could read the book through before you start, to give you an overview.

Whatever method you employ, keep sight of the flow diagram and use it to guide you through the different stages of tackling a project.

Tackling the project—a summarised step-by-step guide

(At all stages seek help and guidance from your supervisor as necessary)

PROJECT BEGINS

Task	Refs (sections)	Outcome
Initial Preparation	(mainly chapters 1 and 2)	
Make sure you know what the project is about and what is expected of you and your supervisor		
Choose a project	1.1, 1.2, 1.3 and 1.4	
	1.5	
Familiarise yourself with the background to and the reasons for the particular project	1.3, 1.4, all of chapter 7	
Appreciate the place of investigations within a project	2.1, 2.2 and 2.3	INVESTIGATION CHOSEN
Break down the project into a set of possible investigations	2.4	
Choose an investigation to work on	2.4, 5.2, 5.3 and 5.5	
Check if another investigation can be carried out at the same time	2.1	
Formulate the aim of the investigation precisely	2.5	
Check whether the aim can be extended to increase the number of results with minimal effort	2.6	
Plan to read the literature to help with the next stages if necessary	All of chapter 7	

Task	Refs (sections)
Design the Investigation	(mainly chapters 3 and 6)
Identify the steps necessary in an investigation	3.2

another investigation

| Plan how the results are to be analysed | 3.4, all of chapter 6 |

Do the Practical Work | (mainly chapters 4, 5 and 7)

Start as soon as the design and planning are ready	4.1
Decide what preliminary investigations are necessary and possible and do them	4.2 and 4.4
Plan and organise the full investigation	4.3
Do the practical work for the full investigation	4.4, 4.5 and 5.6
Make sure you are reading the literature	All of chapter 7
Analyse the results	5.7, all of chapter 6
Interpret the results	5.7
Decide whether further investigations are to be carried out	5.3, 5.4 and 5.8

RESULTS OBTAINED FROM PRACTICAL WORK

Write the Project Report | (mainly chapter 8)

Plan the report overall	8.1, 8.2, 8.3 and 8.4
Collect together Results summaries (conclusions)	5.7
Write the Results and Methods section	8.5 and 8.6
Write up the Introduction and Discussion	8.7, 8.8 and 8.9
Write up the remaining parts of the report	8.10

PROJECT REPORT WRITTEN UP

PROJECT FINISHED

1

General Background to a Biology Project

This chapter will help you to:

Understand what a project is and why it is important

Choose a project

Recognise what is expected of you and your supervisor

Appreciate how a project develops through investigations

1.1   General background

A biology practical project is a practical investigation of a defined aspect of an organism's biology, carried out by the student, and guided by the supervisor, in a limited amount of time determined by the course being followed (for example, 60 hours in a BTEC HND/C project unit). The particular piece of biology being investigated is shown by the title of the project which also indicates the subject area, the approach, and often identifies the overall aim of the project.

The approach taken can vary. It can be one in which a technique, method or procedure is developed ('technical' approach), or it can take an experimental approach in which the experimenter alters an

Table 1.1

A list of project titles in which the overall approach and subject areas have been identified

Overall approach	Project title	Subject area
'Observational'	Seasonal fluctuations in heavy metal levels in two species of upmoorland grass *Molinia caerulea* and *Deschampsia flexura*	Ecology (Bio)Chemistry
	Circadian variation in body temperature and pulse rate	Animal physiology
'Experimental'	The effect of temperature on the level of cyanogenesis and release of hydrogen cyanide from intact plants of white clover and cabbage	Plant physiology
	An experiment into differences in behaviour of resistant and susceptible strains of *Sitophilus granarius* to malathion	Animal behaviour
'Technical'	The use of a microcomputer system for the analysis of intracranial pressure waveforms	Computing Animal physiology
	An investigation into a novel method of detecting essential fatty acid deficiency	Biochemistry
Mixed Observational/ technical	Observation and isolation of golgi bodies from marine algae: *F. serratus*; *F. vesiculosus*; and *P. crispata*	Cell and plant biology
Experimental/ technical	The use of indirect agglutination to detect humoral immune responses to haptenated self proteins in mice	Immunology

experimental factor (often called the variable) in a controlled way to test the organism's response to that factor applied ('experimental' approach), or it can be one in which the aim is to define certain characteristics of an organism (and perhaps make comparisons with others) but not in response to an applied factor (the 'observational' approach), or the approach could be a combination of these ('mixed' approach). Table 1.1 gives a list of project titles and identifies the subject area of the project and the approach to be taken.

The project is an important part of the education and training of biologists as it gives first-hand experience of some of the procedures and methods of original research work (albeit in a small, but nevertheless significant, way). It therefore places emphasis on a sound scientific approach which involves a number of varied activities from careful reading of the literature, good design, careful planning and organisation of the practical work, sound laboratory practice, critical analysis and interpretation of results, to good presentation of the work in a project report.

Since the project is such an integral part of a course and essential training for biologists, it is a time when you, the student, will be learning, and a component of the course where your performance will be assessed. It is therefore reasonable to expect that mistakes will be made as you proceed with the project, so you must not expect everything to be perfect and trouble-free. When you finish the project you will probably wish you could do it again as it would be so much better second time around. FINE! that means you have learned. However, knowing it is to be assessed you will no doubt wish to perform as well as possible to obtain a good mark at the end. Assessment schemes do vary but basically your work should be sound scientifically, produce good results and be well presented and discussed in a final project report. This book offers advice and guidance on how to achieve these aims. The information is based on the common mistakes made by students, the requests for help made by each new group of students, and the comments and criticisms expressed by supervisors and examiners, a selection of which is shown in table 1.2.

Table 1.2

Examples of comments and criticisms made by supervisors and examiners in making assessments

---

'An excellent piece of well thought out and executed work. The practical competence demanded in these techniques is high and this candidate achieved this.'

'It is evident that the overall organisation of the experimental work was poor. The approach is naive and piecemeal. No attempt made to design a programme where quantitative results could be obtained.'

(*continued*)

'Project suffered because the student did not or refused to consult the supervisor.'

'The student did not demonstrate an understanding of the area investigated.'

'The report shows determination, initiative and interest but it is most regrettable that there are so many grammatical errors and mis-spellings.'

'The literature search was inadequate with most of the references being more than seven years old, and scant evidence of having read the original literature.'

'The introduction was relevant and related to the aims of the investigation which were clearly stated.'

'Basic methodological points missed, e.g. inclusion of relevant controls, establishing reproducibility of responses and parameters of stimulation, equipment not calibrated.'

'Early failure of the method did not deter the student but caused him to direct his efforts to careful analysis of the methodology.'

'Planning and execution excellent.'

'Failed to include a text in the results section indicating that the student does not regularly read experimental papers otherwise this would not be done.'

'Presented traces not very convincing.'

'The results are discussed critically with reference to the literature whose source is indicated clearly, and discussed with an obvious awareness of the limitations of the methods used.'

'Little evaluation of own results.'

'A very well presented project report. The student has balanced the various aspects of the project to produce a professional piece of work.'

1.2   The development of a project

The title of a project is normally put forward by a lecturer who will act as supervisor for that project. The title is either the overall aim or a reflection of it, and should be proposed on the basis of ideas which have arisen from careful analysis of either unpublished results or results reported in the scientific literature. The aim of the project should therefore have a sound base with very clear reasons why it was proposed. It should also be possible for a student to carry out the project, and obtain results with the time and facilities available to the department. (The aim therefore should be matched to the department and course.)

The overall aim of the project (as indicated in the project title) is often far too broad to provide the basis for designing and planning practical work. Instead it needs to be broken down into a set of investigations, each with a very precise and well defined aim, to study one specific aspect.

> *Example: Project title 'Amino acids in germinating seeds'*
>
> Specific investigations could include: a determination of total amino acid content of seeds germinated for one day; a comparison of amino acid levels in one-day-old germinating seedlings of different varieties of pea; a determination of amino acid levels over a period of germination in different varieties of seeds; the isolation and identification of amino acids in germinating seeds, etc.

Once the aim of one investigation has been clarified the investigation can be executed and results obtained. As results are obtained, new ideas emerge for other investigations and so the process continues almost never ending. The following flow diagram summarises the process:

PROJECT TITLE
↓
OVERALL AIM OF THE PROJECT
↓
→ SPECIFIC INVESTIGATION IDENTIFIED
↓
AIM OF INVESTIGATION FORMULATED PRECISELY
↓
INVESTIGATION DESIGNED AROUND THE AIM
↓
PRACTICAL WORK PLANNED, ORGANISED AND EXECUTED ACCORDING TO THE DESIGN
↓
RESULTS OBTAINED
↓
RESULTS ANALYSED
↓
RESULTS INTERPRETED
↓
NEW IDEAS FORMULATED FOR INVESTIGATIONS

Flow diagram to show the progression of a project

The key to a project then lies in the investigation and it is this which provides the basis for the practical work. Chapter 2 examines general aspects of investigations in more detail including formulating the aim, chapter 3 considers design, chapter 4 the planning and execution of the investigation, and chapter 5 deals with the results of the investigation. In order to proceed with a project therefore, you will have to first identify an investigation to start on.

1.3 What is expected of the student

What is expected of the student will depend to a large extent on the course being followed. In some courses, as for example in BTEC HND/C courses, there are a number of well-defined objectives which should be achieved, whereas in other courses, for example in some degree courses, the objectives are not laid down so clearly. However a few generalisations can be made. Since the project is an integral part of a course it will be expected that you, the student, will complete the project satisfactorily and in so doing learn from it, developing particular skills and gaining experience and insight into the procedures and methods of original scientific investigation. Listed below are some fundamental and more particular things that will be expected from you.

WHAT WILL BE EXPECTED OF YOU IN THE PROJECT

A. *PLANNING, ORGANISING AND DOING THE PRACTICAL WORK*—This will be done in relation to the overall aims and title of the project.

B. *ATTENDANCE AT PRACTICAL SESSIONS*—You will be expected to work on your own with diligence and commitment to the project, in an organised, intelligent and sensible manner.

C. *SOUND SCIENTIFIC PRACTICE*—In all its aspects from the design and planning to the analysis and interpretation of the results and presentation of the project in a report.

D. *RESULTS*—On the whole, good results will be expected in the sense that they have been produced from sound scientific investigations. The number expected, and amount of verification, will depend very much on the nature of the project. It will be expected that they are analysed and interpreted.

E. *INITIATIVE*—This will be expected in the various areas of the project from ideas relating to the project's development to interpreting results, solving problems and overcoming difficulties.

F. *DISCUSSION WITH THE SUPERVISOR*—You may have to approach your supervisor.

G. *A PROJECT REPORT*—Written up according to standard scientific style and format.

H. *A KNOWLEDGE AND UNDERSTANDING OF THE AIMS AND BACKGROUND TO THE PROJECT*—Therefore, too, a knowledge and understanding of the related literature.

I. *A CRITICAL ASSESSMENT*—Of both your own work and of the published work in the literature in relation to it.

J. *IMPROVEMENT WITH EXPERIENCE*—As you practise the techniques you will improve.

Since the project is part of a course it will be assessed. The assessment (the final mark given) will depend on what is expected of the student and therefore on the major areas shown in the above list. On the whole, such assessment is made by the supervisor and examiners from how the student performs in the lab, the quality of the written report and from the general attitude and approach conveyed and initiative and insight shown in discussions with the supervisor. There may also be an oral examination ('*viva voce*'). In a number of instances the assessment may be based almost entirely on what is finally presented in the project report since it will indicate the quality of the thinking, methods, interpretation and understanding of the work by the student.

If you are not sure what is expected in detail, check with your supervisor and find out what assessment scheme is to be applied to the project. This will give you a fair idea of what the lecturers expect. Meanwhile, use the list given above as a very basic list of what you will have to do and try to keep to it. This book gives general advice on how to attain what is listed.

1.4   The supervisor's role in the project

What exact role the supervisor is expected to play will vary according to the course being followed, the department and the institution. Overall it is true to say that the supervisor plays an advisory and guiding role since the project is regarded as the student's work and therefore an important part of his education and training from which he must learn. The supervisor is therefore a person to whom you can turn for help and advice on the various aspects of project work by discussing points with him. You should not expect the supervisor to provide all the answers, solve all the problems for you, propose all the ideas or work alongside you in the lab or field. You can however expect him to be willing to discuss points with you, either as you need help or on a regular basis, and in so doing encourage you to think for yourself and help you to avoid making serious errors.

Initially a supervisor should ensure that you can get started on the project, and on a project which can produce results.

INITIAL HELP YOU CAN EXPECT FROM YOUR SUPERVISOR

A. *PROPOSE A PROJECT*—It should have a clearly formulated aim and have a background from which the reasons for doing the project are clear, and for which it is possible to obtain results in the time and with the facilities available, given reasonable diligence on the part of the student.

B. *PROVIDE YOU WITH KEY PAPERS OR REFERENCES*—You can then read these to become familiar with the background to the project.

C. *DISCUSS FULLY WITH YOU POSSIBLE LINES OF INVESTIGATION WITHIN THE PROJECT*—He should help you to identify and decide on one to start.

D. *MAKE ARRANGEMENTS FOR YOU TO BECOME FAMILIAR WITH ANY TECHNIQUES, METHODS, INSTRUMENTS ETC. THAT ARE NEEDED FOR YOUR PROJECT*

E. *CONFIRM THAT YOU KNOW WHAT IS EXPECTED OF YOU*—This includes ensuring you are aware of the assessment procedure for the project.

Once this basic information has been sorted out and you begin the practical work, the supervisor should stand back and 'leave you to it' so that you can learn and develop by doing the work for yourself. As the project then progresses the supervisor should provide the following services, though the onus may be on you to obtain help at the appropriate time.

HELP YOU CAN EXPECT FROM THE SUPERVISOR WHEN THE PROJECT IS UNDERWAY

A. *DISCUSS THE DEVELOPMENT OF THE PROJECT*—All its aspects should be open to discussion (you could suggest various ideas and see what the supervisor has to say).

B. *CHECK THAT YOU ARE PROGRESSING SATISFACTORILY*—(You could show the supervisor what exactly you are doing and ask him to assess your progress).

C. *DISCUSS AND INTERPRET THE RESULTS WITH YOU*—You must make sure that the results are in a comprehensible form (and not, for example, your original and raw readings). It is very difficult to discuss results sensibly if they have not been drawn together in some way and presented clearly.

D. *READ CRITICALLY A DRAFT COPY OF THE PROJECT REPORT*—You can work from this to produce the final report.

If your supervisor is not providing the help identified above do seek help from him as these points identify what can be reasonably expected as a minimum from your supervisor. If for some reason you are still unable to get the advice or help you need from your supervisor seek help from some other informed person with experience, for example another lecturer in the Department. Some supervisors are extremely good and will provide you with help automatically, others will be poor and have little to do with you. Some may be able to give you a lot of time whereas others who are very busy may have little time available. On the whole you can expect that lecturers actively involved in research in a similar area to that of your project will be interested in what you are doing. Supervisors who tell you to sort out a project, or find the background for yourself from scratch probably have not thought the project through very well, and busy senior departmental staff, or lecturers supervising a large number of projects will probably not be able to give you a lot of time even if they wanted to, though they will probably be interested.

1.5 Choosing a project

It is very likely that you will be given a choice of project from a number available in the department which reflect the interests, research and specialisms of the members of staff. From the projects available you will then have to make a choice (it is probably unlikely that you will be obliged to come up with a project title of your own). In making a choice of project you will be trying to choose one which suits you from many angles—your interests, temperament, needs, aptitudes, dislikes and likes etc. In reality it is unlikely that the project you finally do will suit you in all possible ways. You will therefore have to compromise to a large or small extent. It does not often matter too much that you will have to compromise, as once you get involved in a project it becomes more interesting than it might have seemed at the outset. However, in compromising it is important to recognise that since the project is part of your course you will have to complete it satisfactorily. It makes sense therefore to avoid a project which really does not suit you, since if it does not suit you in a lot of ways, you will probably find it difficult to give the commitment necessary and persevere in the face of difficulties which will inevitably occur.

In choosing a project therefore, take note of the following.

FACTORS TO TAKE INTO ACCOUNT WHEN CHOOSING A PROJECT

A. *YOU DO NOT INSTANTLY DISLIKE THE SUBJECT MATTER OR APPROACH OF THE PROJECT*—As doing the project does require a good deal of commitment, do not do something you really hate or dislike, for example you might hate field work, find it difficult to work with live animals, not be technically-minded, imaginative or inventive. If you are aware of such dislikes, make sure the project does not involve these to a large extent.

B. *THE TECHNIQUES ARE NOT COMPLETELY BEYOND WHAT YOU ARE CAPABLE OF*—You probably have a good idea about how technically-minded you are and your general lab skills. If you find, for example, that you have great difficulty understanding how to use complex electronic apparatus, or have a mental block about making up different types of solutions, do not choose a project which has these as an integral part unless you specifically wish to use the project to improve on such weaknesses or lack of knowledge and understanding.

C. *YOU AND YOUR SUPERVISOR DO NOT DISLIKE ONE ANOTHER INTENSELY*—The supervisor can be a tremendous help in guiding you carefully through the project from the basic idea and aim of the project through to analysing and interpreting the results and writing the report. The supervisor can be a real stimulus in making the project an enjoyable and worthwhile experience in which you learn a lot, and in helping you to perform better and thus obtain a good mark. If therefore you do not get on you will probably find it difficult to approach the supervisor when you need help, and the supervisor may find it difficult to be generous in his support.

D. *YOU ARE LIKELY TO OBTAIN RESULTS*—It *is* important to ensure that there is a high chance that results can be obtained if you put in the expected effort and commitment. Results are generally expected from the project unless things go wrong through no fault of your own. Results add interest for you in the project, they give you a greater opportunity to demonstrate other abilities (for example, critical interpretation) and make it easier to write the report. (Chapter 5 considers results more fully, including what a result is.) Basically, the chances of obtaining results decrease as the complexity of the methods increase and if the techniques are new for the organism investigated. It therefore follows that you are more likely to obtain results if the techniques are well established and well tried for your organism, and if the methods are relatively simple overall. Hopefully the supervisor will have proposed a project in which it is reasonable to expect results with the time and facilities available and for which he has a clear understanding of what can be achieved which can be conveyed to you.

When therefore making a choice from those available, take account of the points above and then discard all those projects which you do not like because of the subject matter, approach or supervisor. If you dislike them all, choose ones you dislike the least. With the titles you have left, delay your final choice of project until you have more information about what is involved (ideally by discussion with the supervisor, though you may be able to deduce what you need from the title). To make or confirm a final choice you will have to have a very clear understanding of what the project is aiming to achieve, the reasons why it was proposed and a realistic appreciation of the way in which the project is likely to proceed to obtain results. The kinds of things that should be established are: what kind of approach will be taken (technical, experimental, observational—see table 1.1); what exactly is the project about; what methods are to be used; are the methods well established or will they have to be developed; are the methods complex or new for the material being worked on; are results expected quickly in the short term or will there be some delay (for example, plants may have to be grown for a period under experimental conditions before results can be obtained); how much does the project depend on apparatus or instruments of limited availability (could that be a problem); are difficulties likely to arise—what kind and is it likely that they can be overcome easily; are there any difficulties in obtaining the biological material, etc.?

With this kind of information about each project, make your final choice. Try to match the project you choose to your own interests and needs as well as possible and take account of the points listed previously. With the project chosen you can now begin to consider what specific investigation to carry out and refer to chapter 2.

Having considered all these various points it might seem to you that you will have to spend a great deal of time choosing a suitable project from those available. However, in reality, it is likely that only a very limited number suit your interests and abilities so that your real choice of projects is quite limited. It is important to try to make your choice quickly, or at least identify the 'possibilities', and to approach the supervisor concerned to avoid being disappointed because someone else has chosen the project before you.

If you do not have a choice of project and are not keen on what you have been given, take comfort in the fact that a project can generally become interesting if you do try to get involved with it. Not only that, but you may be able to take an approach which matches your interests and needs more closely when you are planning and designing specific investigations.

If you are given complete freedom to come up with your own project the points given above are important in making a choice, but in addition there are many other factors which have to be taken into account relating to the background of the project and the feasibility of the project in the Department and for your course. My advice would be not to proceed until you have obtained informed opinion from someone with interests and experience in the field you are concerned with. Their experience is greater than yours and they will have a deeper appreciation of what can be reasonably achieved and so can advise and direct you accordingly. I think it is generally a mistake to start and work completely on your own.

2

The Investigation

This chapter will help you to:

Understand how the investigation fits into the project overall

Recognise the basic characteristics of an investigation

Identify the two categories of investigation

Choose an investigation to start with

Formulate the aim of an investigation precisely

Extend an aim to increase the number of results

2.1   The place of the investigation within a project

It is generally necessary to divide the project into a set of *investigations* where each has a precise aim to study one specific, well-defined aspect of the project. The practical work is then designed and carried out to attempt to satisfy the specific aim of an investigation. The investigations together form a coherent group interrelated by the overall aim or title of the project.

Ideas for investigations arise from examining the title, or overall aim, of the project and by understanding the background to it. In some cases it may be possible to identify a number of investigations. This generally occurs when the title or aim of the project is stated in broad terms, such as 'The analysis of human breast milk'. The detailed investigation is not identified and a number of approaches could be taken, for example, the identification and relative numbers of cells; determination of fat and protein content over two weeks after the birth of the child; or analysis of antibody activity. Alternatively, the title may be so specific that one obvious investigation is evident; for example, 'An investigation into the effects of exogenous application of the growth regulator, giberillic acid, on the extension of *Pisum sativum* root segments, 10 mm in length, with intact root tips'. In either case, there is always an investigation from which to proceed. Other investigations then arise from it which can occur in series or in parallel. Investigations which occur in series depend on the results of an investigation previously carried out in the project. They cannot therefore be carried out until one investigation is completed since the design of subsequent investigations depends on the results of the first. Parallel investigations on the other hand do not depend on the results of former work carried out during the project, so it is possible they can be carried out concurrently. Such investigations can make good use of your time. They can be particularly useful if one of the investigations is very original, with many uncertainties about whether results can be obtained. Another investigation could be carried out in parallel in which there was a high degree of certainty that results would be obtained. An example of how investigations can occur in series and in parallel is given opposite.

The number of investigations which are therefore evident at the outset of the project will depend on the precision of the project title or overall aim. If there are a number which can be proposed a choice will have to be made (see section 2.4). The number of investigations which can then actually be carried out in a project will depend on the nature of the investigations themselves. It is always expected that at least one is finished, and in some projects where the methods are simple and results easily obtained, more completed investigations would be expected.

The investigation therefore is the basic unit of the project from

Example: Project title 'Feeding preferences in slugs'

Parallel investigations could be: testing a slug's preference for different varieties of potato; different varieties of brassicas; different ages of brassica material; working on the same plant material but with different slug species; or doing investigations which cover different time periods for testing the preference. An investigation which could follow after some results have been obtained could be to test whether the slugs showed any preference when given a choice between the preferred variety of brassica or potato (determined from the previous work), or to try to determine the chemical basis of the preference by, for example, making extracts of the plants and trying to identify the 'active' constituent(s). Another in series investigation might be to record from the sensory nerves to see if the nerve activity could be related to the preferences shown or chemicals extracted.

which practical work is designed, planned and carried out, and results analysed and interpreted. Every effort should be made to ensure that the investigations are successful, in the sense that verifiable and meaningful results are produced, since the success of the project depends ultimately on the success of the component investigations.

For all projects it will be necessary to carry out the following.

COMMON INITIAL STEPS IN A PROJECT

A. *CHOOSE AN INVESTIGATION TO START ON*—Section 2.4 provides some suggestions of what could be usefully considered.

B. *DEFINE THE AIM OF THE INVESTIGATION PRECISELY*—Section 2.5 shows you how to do this.

C. *CONSIDER EXPANDING THE AIM FOR INCREASING THE NUMBER OF RESULTS*—Section 2.6 shows you ways of doing this.

It will be helpful for you in your project to have a good understanding of what an investigation is and its basic characteristics. Sections 2.2 and 2.3 consider these aspects and will help you in choosing an investigation wisely and in the later planning and designing of it. Reference to these sections is therefore recommended.

2.2   Basic characteristics of investigations

In all investigations which make up a project, observations, which could be measurements, are made on some aspect of an organism's biology. The precise characteristic on which observations are made is the variable and can include such things as rate of K^+ uptake, population size, amount of protein etc. The variable is thus what is actually observed and the observations on the variable are the results.

Example

In a project concerned with studying toxicity of nitrite on certain microorganisms, the variable observed was the number of cells remaining after the nitrite had been applied (used as a measure of toxicity). Similarly, in a behavioural investigation on motivation, the variable was the time the animal spent trying to overcome a particular difficulty to satisfy the need which was motivating it. In another instance, in a study of photosynthesis, it was the rate of production of CO_2 which was observed and therefore the variable which was used as a measure of photosynthesis.

The observations made can be described as being quantitative or qualitative. Quantitative observations are ones where measurements are made in the form of numbers of some kind on the variable. A measurement scale is therefore used (nominal, ordinal, interval or ratio, see section 3.4, table 3.1) such as weight, length, absorbance, or number of organisms having a particular characteristic. An instrument usually has to be employed for making the measurement, such as a balance, a ruler or a spectrophotometer. Qualitative observations, on the other hand, are ones which are not described by numbers as a result of making measurements. Instead, they take the form of written descriptions, diagrams or photographs; for example, a muscle cell structure could be described in words, illustrated by diagrams or shown by photomicrographs. (The muscle cell could also be described quantitatively by, for example, giving the dimensions of the cell, or working out the number of mitochondria present, or grouping the muscle cells into categories which describe a particular characteristic.) Both types of observations are useful but it is advisable to try to use quantitative measurements whenever possible as they are more readily open to a range of methods of analysis (such as giving a mean value), make it easier to make precise comparisons with other data and are often more precise than qualitative measures.

In order to collect data on the variable the investigation has to be carried out in two distinct stages:

FIRST STAGE—the material has to be obtained in a form where the variable can be observed/measured.

SECOND STAGE—the data on the variable has to be collected by making appropriate measurements.

For each of these stages, techniques will have to be employed. The examples below identify the two stages in different investigations, and the basic method to be used for each.

Examples

Project title 'The effect of zinc on the oxygen dissociation curve of rat blood'
FIRST STAGE—the rat blood has to be obtained. A hypodermic syringe was used to remove the blood from the tail.
SECOND STAGE—the astrup, an instrument for obtaining an oxygen dissociation curve was used on the rat blood.
(Both stages are relatively easy and routine.)

Project title 'The ultrastructure of isolated mitochondria in germinating grass (Agrostis sp.) seedlings'
FIRST STAGE—seeds were obtained from the supplier, and were germinated. Routine fractionation procedures were used for obtaining the mitochondria.
SECOND STAGE—the ultrastructure of the isolated mitochondria was determined using the electron microscope.
(Both stages are routine but time-consuming.)

Project title 'A determination of the K_m of lactate dehydrogenase from pig heart'
FIRST STAGE—the enzyme has to be obtained from the heart which was supplied by the abattoir. Various procedures are involved to extract, purify and crystallise the enzyme.
SECOND STAGE—spectrophotometry can be used to determine the K_m.
(The first stage may be time-consuming and full of problems not yet identified, whereas the second stage is routine. The project may ultimately concentrate on developing methods for extracting the enzyme in reasonably pure form and in sufficient quantities to investigate.)

2.3   Categories of investigation in a project

No matter what approach is taken in a project, or its subject, the investigations within it fall into one of two categories. Each has a different approach and inherent problems and difficulties which have to be taken into account in the design.

CATEGORY A—Investigations in which the investigator does not alter an experimental treatment in a deliberate and controlled way to test its specific effect on the variable, but collects data on individuals from a clearly defined population.

Data are collected on an identified variable, from individuals in a particular population or group. The investigator does not impose any controlled treatments or experimental conditions on the organism though the population from which the organism is drawn may be described in precise terms—for example, population of diseased pea plants (of a particular defined variety) from clay soil in Norfolk. The major aim of such an investigation is to define some characteristic of the group or population by making observations/measurements on variable(s). Additionally, the aim may extend to making comparisons with another group or population. Examples are shown in table 2.1.

Table 2.1
Examples of investigations in category A

	Variable	*Population or group*
1. A comparison of	the ion composition of body fluids	of two species of snail obtained from the wild in summer
2.	The ultrastructure of	*Chlorella* (a green alga)
3. A study of the correlation between	shell height and width	of dog whelks from exposed and sheltered shores
4. A determination of	the quantity of glycogen	in rat liver
5. The identification of	species	of microorganisms in restaurant kitchens
6. A comparison of	lead levels	of cereal plants alongside and separate from motorways

In such investigations the common difficulty is to ensure that the individuals on which observations were made, and from which data on the variable are collected, represent a homogeneous definable group or population. In addition, there are often difficulties concerned with ensuring that a large enough, representative sample of individuals is obtained.

CATEGORY B—Investigations in which a specified experimental treatment is applied in a controlled way to individuals of a defined population and data collected in response to that treatment.

Once again, observations are made, and data therefore collected, on a defined variable from a group or population, *but* in response to an imposed experimental condition applied in a controlled way by the experimenter. The main aim is to determine whether the treatment applied has any effect on the variable studied and if so the nature of that effect. Examples of investigations in this category are given in table 2.2.

Table 2.2
Examples of investigations in category B

	Experimental treatment	Variable	Population or group
1. The effect of	the fungicide 'Prochloraz'	on the *in vitro* mycelial development of	the pathogen *Rizoctonia cerealis*
2. A study of the effects of	postural change on	the limb blood flow of	humans
3. The effect of	temperature	on the germination of	*Molinia sp.* seeds
4. The effect of	four different soils	on the root length of	tomato seedlings
5. The effect of	three extraction procedures	on the activity of	lactate dehydrogenase

The major problem is to design an investigation to enable valid conclusions to be drawn about whether the applied treatment has had an effect or not, and if so the details of that effect. To do this it is necessary to make comparisons with organisms which have either not received the treatment (controls) or which have received a different level of treatment, but which are similar as far as possible in all other respects. The difficulty therefore is in keeping all uncontrolled sources of error to a minimum and ensuring that conditions are as similar as possible for each of the groups being compared. However, the advantage of this type of investigation is that the control of the experimental conditions is largely with the experimenter, so he may be able to reduce other sources of error, or possible causes of differences in response, to a minimum.

Example

Suppose an investigator was interested in determining what changes occurred in an animal infected by a particular pathogenic organism. An approach consistent with category B would be to obtain two similar samples of organisms from the same population and to deliberately infect one group. (The choice of which would be infected would be random.) The other group would be treated in as similar a way as possible but not infected, and results obtained. The investigator controls the investigation very precisely to determine what changes were due to the infection. In an investigation of category A type, the investigator would obtain a sample of infected animals and a sample of non-infected animals from a population and determine what differences were apparent. He would not be able to conclude what differences were *due* to the infecting organisms but what differences *occurred in* an organism with the infection.

2.4   Choosing an investigation

Having to choose an investigation to start on generally occurs when the project title is broad. The best thing to do is to start by looking at the overall aim and background of the project and to begin to list possible investigations. Initially, this might take the form of asking a number of questions.

Example

In a project concerned with the growth of fern gametophytes, one direction may be to examine various aspects of cell activity during growth of the protonema, and questions which you may ask could be: which cells divide and when? which cells expand in which direction and for how long? what is the effect of red, blue and white light on growth? how do the divisions giving rise to rhizoids consistently differ from those producing only chlorocytes (cells containing large chloroplasts but no proplastids) etc.?

You then need to visualise what investigation could be carried out to answer some of the questions and decide whether a single investigation may be able to answer a number of questions at one time. Formulate the aim of the investigation very clearly as recommended in section 2.5. You can then look at the aims of the investigations you have come up with and choose one to start on.

In choosing an investigation, make sure that it is closely related to the overall aim of the project, that it will work, that you can see what developments could occur from it and take the following points into account.

FACTORS TO TAKE ACCOUNT OF IN CHOOSING AN INVESTIGATION

A. *FIND OUT HOW MUCH TIME IS AVAILABLE FOR THE PROJECT AND HOW IT WILL BE ORGANISED*—Any choice of investigation should take this into account so that it can be completed within the time constraints.

B. *ENSURE THE INVESTIGATION MAKES SENSE BIOLOGICALLY*—Make sure that there is a good biological reason for the investigation and that there is a clear background to it. In other words, try to justify your investigation in biological terms and in relation to the overall aim.

Example

A student thought it a good idea to test the effect of varying salinities on the locomotory behaviour of winkles, but he was not sure why it was a good idea except that the animal was a marine organism and 'marine equals salt'. Such a study could be justified if one could argue that the animal may be subjected to varying salinities in its normal environment, or if some aspect of tolerance was to be investigated. When the justification is clear the investigation can have an aim and design which makes some biological sense.

Though it is, of course, possible to carry out investigations which have little justification or relevance in biological terms, they are much less satisfactory and will produce difficulties in interpreting results and in writing up unless a justification is found (or manoeuvred) in hindsight.

C. *ASSESS WHETHER THE INVESTIGATION IS FEASIBLE*—Identify in a general way what will have to be done to prepare the organism for obtaining data and then what has to be done to collect the results; that is, identify the two stages of an investigation as shown in section 2.2. Determine whether these can be carried out and if they cannot, or present unsurmountable difficulties, discard the investigation.

The investigation is feasible if you can obtain the organisms, set them up under the correct conditions for recording data, maintain experimental conditions, and are able to collect results.

Example

There is little point considering an investigation in which daily measurements have to be taken over two weeks when there is no access to the organisms at weekends, or if one-week-old seedlings were to be used but it was almost impossible to germinate the seeds quickly, or if you need an instrument for setting up experimental conditions which is not available.

D. *ENSURE THAT OTHER INVESTIGATIONS CAN DEVELOP FROM IT*—Look at the investigation and try to anticipate alternative outcomes (results) from it, for example, that you do or do not get a response. See then where you could go next in each situation. If you cannot see what can be done next, you will not be able to carry out a series of related investigations. Also try to see what related investigations could be carried out in parallel. If you see no development of the project into other investigations it is best not to include the investigation in your list of choices.

The final choice of investigation should take these factors into account. Having decided on an investigation, formulate the aim very precisely as recommended in section 2.5 below, and consider further whether other parallel investigations could be carried out at the same time (thus maximising the use of your time). Decide, also, whether alternative variables could be investigated. With the aim clearly defined you will be ready to design your investigation as recommended in chapter 3.

2.5   Formulating the aim of an investigation precisely

In trying to define a useful and worthwhile aim around which an investigation can be planned and organised, it is helpful to try to ensure that the aim has the following components to it. An aim with such a detailed formulation is then the working aim for the investigation.

COMPONENTS TO BE INCLUDED IN THE AIM OF AN INVESTIGATION

A. *THE REASON(S)*—For doing the investigation (or the point of it); for example, to determine, to test, to compare, to estimate, to describe, or to distinguish between.

B. *THE BIOLOGICAL MATERIAL*—To indicate exactly what the observations will be made on. This means identifying the population and part of the organism investigated, such as liver cells of the adult laboratory rat (Wistar strain), or the gastocnemius muscle of the frog *Rana pipiens*.

C. *THE VARIABLE(S)*—To be observed or measured; such as the length of the longest root, the strength of muscle contraction, the K_m, the rate of K^+ uptake, the percentage cover, or the degree of inhibition.

D. *THE EXPERIMENTAL FACTOR(S)*—This is used as a general term for categorising similar types of treatment, such as locality, age, species and pH. Levels of treatments applied (for example, species A, B, C. . . etc.) are usefully indicated as well.

E. *ADDITIONAL INFORMATION ON CONDITIONS OF THE INVESTIGATION*—It might be useful to include other information to clarify the aim like, for example, the technique to be employed, additional characteristics of the organisms (such as the fact that they were starved, or were one sex), or periods when recordings were made (for example, summer of 1986).

When including the above components in the aim, be as precise as you can. Since the aim will provide the basis for the design, it should convey as much specific information as possible; for example, give the range of treatment levels, if possible, that is, pH from 5 to 10, rather than just stating a range of pH, or identify what exact parameters of growth were to be measured rather than just stating growth. Tables 2.3 and 2.4 show examples of the aims of investigations and include the components listed above. The examples are classified into the two categories identified in section 2.3.

Clarifying the reason(s) for an investigation is important, as it directs the investigation and also determines what method of analysis of results will be appropriate. On the whole, the reasons for an investigation fall into one of two categories which reflect the type of investigation to be carried out.

Examples: The precise reasons for an investigation within the two categories

Category A—To establish some detail of a variable
Here the main aim is to determine certain characteristics of a variable of an organism. The precise reasons may, for example, be to: define or determine that characteristic; to define then compare with another population; or to test for correlation between different variables from the same individual; or to compare correlation coefficients for two variables of two distinct populations; or to compare the data with a theoretical distribution, such as to test for randomness, etc.

Category B—To monitor the effect of an experimental factor on a variable
Here the main aim is to determine what effect the experimental treatment has had. The precise reasons may be to: determine whether there is a significant difference between the control and the experimental group; to identify if there is a significant difference between different levels of treatments; to identify if there is a trend in the response as levels of the treatment alter; to identify a point at which a treatment has an effect (such as a maximum, minimum, significant effect); or to determine the degree of the effect of the treatments, etc.

Table 2.3

Examples of aims of investigations in category A

	Characteristic component		
Reason	*Variable*	*Population*	*Additional information*
To determine and compare	the total protein content (excluding shell)	of hens eggs of three different sizes	grades (2, 3, 4)
To identify	the bacterial pathogens	in *Taxus baccata*	from two sources
To record	the resting membrane potential	of a range of identified neurons of *Helix aspersa*	

Table 2.4

Examples of aims of investigations in category B

	Characteristic component			
Reason	*Experimental treatment(s)*	*Variable*	*Population*	*Additional information*
To compare the effect of	three different light wavelengths (red, green and blue)	on the growth (length and weight)	of larvae of *Pieris Brassica*	
To test the effect of	a field dose of metaldehyde	on the motility	of the oesophagus of the gut of *Arion ater*	*in situ* for animals starved for 2 days
To determine the effect of	freezing and subsequent thawing	on the respiration rate	of isolated mitochondria from rat liver	using the Gilson respirometer

The precise reasons within each category vary and will be influenced by the background of the project; for instance, it may be hypothesised that a certain dose of insecticide was effective, so an investigation was designed to test whether this was an effective dose or not rather than to test the response to a range of doses.

In some cases there may be more than one reason which can be satisfied in one investigation. This is particularly the case if more than one experimental factor is altered or if more than one variable is measured on the same individual. In the former case the aim would be to test the effect of each of the factors on the variable, and possibly the interaction between those factors, and in the latter case the aim may be to define each of the characteristics of the variable and to test for correlation between the variables.

For your investigation, therefore, formulate the aim as clearly as possible according to the recommendations here. Having done that, check in section 2.6 to see if it is possible to obtain more from your investigation with minimal effort, and then proceed with the design as directed in chapter 3.

2.6 Expansion from the aim to increase results (with minimum effort)

2.6.1   Variables

When formulating an aim for an investigation having used the format previously described, it is always worthwhile considering whether there are any other variables which could be usefully measured or observed at the same time, and with minimal additional effort, to add to the variable already planned for.

Example

Suppose a student was testing the effect of acetylcholine on muscle contraction and had decided to measure the strength of contraction. He would apply the acetylcholine, record the contraction and measure the strength. With very little extra effort he could also measure the duration of the contraction, the time delay between application of the acetylcholine, and the start of the contraction, and the time to reach the peak of the contraction. In this example, the student did not have to set up any further experiments but now has measured four variables rather than one, and therefore has four times as much information as he might have had to present in the project report, with very little extra effort.

Observing a number of variables in one investigation has two major effects: it increases the number of results which can be presented, and gives more information about the system under study so that it may enable more useful interpretations of the results to be made. The additional variables may be qualitative or quantitative measurements, using any of the scales of measurement defined in section 3.4, table 3.1 (that is, nominal, ordinal, ratio or interval). Sometimes it is not always possible to think of extra variables to study at the outset of an investigation, but ideas for additional worthwhile variables may emerge as the practical work proceeds and can be noted and taken account of.

There is a danger of measuring *anything*, but this should be avoided. You must apply some common sense to the number and type of variables chosen and try to make sure they are relevant within the context of the investigation; for example, you could ensure that they are related in some way, like aspects of growth, feeding or muscle response (as in the previous example) and try to see where useful correlations between variables could be made. When planning an investigation, it may be possible to study a very large number of variables. For example, to test the effect of a herbicide on a plant, various parameters of growth could be measured (like length and weights of various parts), respiration and photosynthetic rates determined, chlorophyll content measured, and so on. In such a case the best thing to do is compromise. Choose variables in which the same or a simple technique is applied.

Example

Suppose the aim of an investigation was to use flame photometry to determine the level of Na^+ in crab body fluids over a twenty-four-hour period, taking samples at four regular intervals, to see if there was a diurnal cycle of ion levels. With minimum effort, K^+ levels could also be monitored, so the number of results can be sensibly doubled.

2.6.2   Experimental factors (or treatment types)

Having stated that it is worthwhile increasing the number of variables investigated if it requires minimal effort, it is also worthwhile considering whether increasing the number of factors in an investigation will yield useful results. However, the design will be much more complex—the complexity increasing with the number of factors. It is advisable, therefore, to restrict the number of factors altered to a maximum of two, unless there are very clear reasons for doing otherwise. In doing this, you should be absolutely clear what you hope to find out as a result of altering each of the experimental factors, and what methods of analysis will be used to give the information required. The design for a two factor investigation would thus broadly be as shown in table 2.5 where A–E and I–III identify the specific treatments within each experimental factor. It is likely, but not always the case, that a sample of organisms for each of the combined conditions, defined by the treatment levels, would be used.

Table 2.5 Showing the general design for a 'two-factor' investigation

		One factor, such as temperature					
		A	B	C	D	E	etc.
Second factor such as source of organisms	I	Values of the variables measured					
	II	places in the boxes					
	III						

Thus there would be a number of measurements of the variable for each box (that is, IA, IIB, etc.).

Example

Suppose the respiration rate of goldfish was measured in order to study acclimation. The respiration rates would be measured for a sample of goldfish over a range of temperatures for goldfish which had been kept, for a defined period of time, at different temperatures. All other conditions were kept constant as far as possible. The two factors would then be temperature range and pre-treatment condition. The aim of the experimenter would be to:
 (i) determine how the respiration rate varies over the temperature range for each pre-treatment temperature and;
(ii) to make comparisons between the respiration rates recorded over the range of temperatures for each of the pre-treatments.

As the number of factors is altered within an experiment it becomes both difficult to design the experiment so that it is properly controlled, increasingly difficult to analyse the data in any simple way, and hard to understand what the analysis is demonstrating. This is because with so many factors altered the possible interactions between factors begin to increase markedly.

Example

In a behavioural experiment the aim was to determine if there was a significant difference in the activities of two different strains of grain beetle, *Sitophilus granarius*, using an activity monitor. Experiments were carried out at three different temperatures and recordings of activity were made daily over one week.

The design is thus:

| | | | Sitophilis *strain* | |
			Strain A	*Strain B*
Temp. I	Day of week	1 2 3 4 5 etc.	Variable: activity measured on activity monitor	
Temp. II	Day of week	1 2 3 4 5 etc.		
Temp. III	Day of week	1 2 3 4 5 etc.		

The experimenter wishes to find out if and how the activity varies with the three factors. He could test for:

(i) Overall differences in activity between
   strains
   days
   temperatures
(ii) Differences in the activity for different strains at each time and each temperature.
(iii) The pattern of activity over the days or temperatures for the two strains and the nature of those differences.

From this example, you should be able to see that with three factors (as in this case) it becomes quite complicated to consider all the ways the factors could separately or together affect the activity. In addition, a sample of each strain was used. This further complicates the analysis.

Generally, it is easier to design and analyse investigations in which the number of experimental factors is kept to a minimum, and will probably be clearer for you. When the main aim is to determine whether the levels of the factors are having a significant effect on the variable measured, it is usual to use a two-way analysis of variance for the analysis of the data. This technique is useful not only for determining whether each of the factors has a significant effect on the variable measured, but also for determining whether interaction between the factors is occurring. (This basically means that the combination of the two factors produce an unexpected or different pattern of response from when one factor alone is being tested.) The value of multifactor investigations is that they can yield some extremely useful information about the interaction between factors if the design is carefully planned, and can make maximum use of resources which might be limited (for example, limited in terms of time or material available). There are more advanced statistical techniques available for analysing the data with more than two factors. If you find it necessary or particularly desirable to take such an approach you *must* seek advice and help from a competent statistician *before you start* any practical work in order to clarify the design details.

In general, however, it is probably safer and more desirable for you to restrict the number of factors altered in an investigation to two for your project, particularly if it is part of a course, and where time and facilities are limited, and instead to try to increase sensibly the number of variables investigated. If for some reason you do need to increase the number of factors, clarify, before you start the practical, the detailed aims of the multifactor investigation so you know exactly what you hope to find out, and can thus seek help on the best way to analyse data.

2.7   Final preparation before contemplating the design

Before starting the design you should have identified and chosen an investigation to start with, taking account of the factors listed in section 2.4. Make sure that the investigation makes sense biologically and that it is feasible (that is, that you are likely to obtain results and that you can see how it could develop from different kinds of results).

Define the aim very precisely, taking particular care to make sure that it includes the reasons for the investigation, the variables to be observed, the material to be studied, and the experimental factors, if any, to be applied (section 2.5).

Recognise whether the investigation you have chosen is of the category B type where you will be testing the effect of an applied experimental treatment, or of the category A type in which no treatment is applied in a controlled way (section 2.3). If you do this it will help you with the design.

Finally, think about whether it is worthwhile extending the investigation in any way to study more variables or applied treatments in the same investigation (section 2.6). Thinking this through carefully could help you to increase your results with minimum effort. Consider the possibility of using a range of measurement scales for different kinds of data (see section 3.4, table 3.1).

3

Designing an Investigation

This chapter will help you to:

Recognise what questions should be asked and answers found in a design

Breakdown your investigation into its component serial steps

Decide what methods to employ at each step

Take appropriate account of analysing the results

Decide what controls, samples and replication will be necessary

3.1   General aspects

The design of an investigation provides the basis for the practical work and as such is an important stage in the development of a project. A good design is essential for ensuring that the practical work is scientific, produces good results, and makes maximum use of time and facilities. In designing an investigation you are obliged to 'think it through' very thoroughly to ensure a sound scientific approach, and to make decisions relating to the methods to be used for obtaining and analysing results. In formulating a design for an investigation it is first necessary to have a well-defined aim, as advised in section 2.5 of Chapter 2, so that it is clear what variables are to be measured/observed, on what organism, under what conditions and for what reasons. The design can then be carried out in a number of stages to answer the questions listed below. You can proceed by going through stages in the series as listed. However you will find that as the design develops, you will have to refer back and double check on all the stages so that finally the design is coherent and complete. A good practice is then to test the design with preliminary investigations to ensure that results relating to the aim can be obtained. However, this is not always feasible in a short term project (section 4.2).

DECISIONS TO BE MADE IN DESIGNING AN INVESTIGATION

A. *WHAT STEPS ARE NECESSARY IN THE INVESTIGATION*—To make observations/measurements, and obtain results to achieve the aim? (section 3.2)

B. *WHAT METHODS WILL BE USED AT EACH STEP?*—Techniques, apparatus, instruments, procedures. (section 3.3)

C. *WHAT METHODS WILL BE USED TO ANALYSE THE RESULTS? (section 3.4)*

D. *WHAT KINDS OF CONTROLS AND REPLICATION AND/OR SAMPLING ARE NECESSARY*—To verify the results and validate the findings? (section 3.5)

3.2   Identifying the steps in an investigation

There are a number of steps which may have to be carried out in an investigation and which the design has to take account of. Listed below are the possible steps and * identifies those common to all. You will need to identify which are necessary in your investigation.

POSSIBLE STEPS IN AN INVESTIGATION

*A. *PREPARING THE ORGANISM TO A POINT WHERE OBSERVATIONS/MEASUREMENTS CAN BE MADE*
Any number of the following may be involved:

*(i).   *COLLECTION OF THE ORGANISM*—You may have to order the organism from a supplier or collect it from the field.

*(ii).   *MAINTENANCE OF THE ORGANISM*—Once collected the organism will have to be maintained in conditions which will keep it alive and healthy or which will preserve it in a suitable state (that is, some organisms may be preserved immediately they are collected).

(iii).   *PRE-TREATMENT OF THE ORGANISM*—It may be necessary, for example, to starve an animal, grow a plant to a particular stage of development, culture some microorganism on a selective medium or infect an organism.

(iv).   *EXTRACTION OR ISOLATION OF A PART OF THE ORGANISM*—You may need to extract a particular part of an organism or isolate part of it in order to collect results. It is common that there are established procedures for this, such as fractionation for isolating organelles, dissection methods for an organ or tissue, and extraction methods for certain (bio) chemical compounds.

(v).   *MAINTENANCE OF THE ORGANISM IN THE EXTRACTED OR ISOLATED STATE*—Once isolated you will probably need to keep it in conditions which cause minimum disruption to its normal state; for example, if you need to keep it alive the conditions should mimic as closely as possible the conditions it is normally in, or, if preserved at this stage, in a preserving condition which changes it as little as possible for the variable investigated.

(vi).   *COLLECTION OR PREPARATION OF OTHER MATERIAL NEEDED*—You may need to have, for example, feeding material for your investigation, or certain drugs, or extracts of another organism, etc. These have to be collected and kept in a state suitable for the investigation.

*(vii).   *SETTING UP THE ORGANISM, OR PART OF IT, FOR MAKING OBSERVATIONS/MEASUREMENTS*—Material has to be prepared in such a way that the results can be collected. This may mean, for example, attaching an instrument, placing the organism in an instrument, or conditions for measurement, preparing slides for microscopy, etc.

(viii).   *MAINTAINING THE ORGANISM WHILE RESULTS ARE COLLECTED*—The organism may need to remain in a healthy, functional state while results are being collected, and the best conditions have to be found for this. The aim is to try to ensure that the process for the collection of the results itself does not affect the result.

(ix).   *APPLYING THE EXPERIMENTAL CONDITIONS*—If the effect of some treatment has to be observed, then the treatment will need to be applied at some stage. The methods chosen to apply the treatments should cause minimum disruption and normally be controlled.

*B. *COLLECTION OF THE RESULTS*

In all investigations, methods have to be employed to collect the results, that is, obtain data in the form of observations or measurements of the variable. It might, for example, be observing and taking notes, recording with photographs or using an instrument or technique to register and record responses. The method must give data (of the variable) which can be analysed as planned so that the aim will be satisfied; for example, if you are interested in measuring the absorption of a pigment from the ultra-violet to the infra-red range, a spectrophotometer with such a range of wavelengths must be chosen.

   The number of steps in an investigation can vary enormously with different projects. Some have relatively few preparatory procedures, and others many—where the actual collection of data becomes a long way removed from first obtaining the organism.

Examples

1. 'Response of nerve cells of locust to odours'
Locusts have to be ordered, and kept in conditions to maintain them. The animals are left for at least one week after arriving at the lab to settle, nerve ganglia have to be exposed, ganglion bathed in insect saline, electrodes made, the locust fixed rigidly for recording and electrodes placed in a ganglion, animal kept in constant conditions while recordings made, odours tested with adequate controls, records of response shown on oscilloscope and stored on tape and perhaps analysed with aid of computer at the same time.

2. 'Determination of gut content of roach'
Samples of fish collected from identified source, animals taken to lab (and frozen if necessary until they can be examined, then defrosted), animal dissected, gut removed, contents placed in water and examined with the eye and light microscope.

The more steps involved the more complex the overall design becomes, and can be complicated further by procedures which have to be applied at each step. With more steps involved there are theoretically more places at which things can go wrong. Careful planning of each step is extremely important to try to avoid difficulties and to increase the chances of obtaining good results.

3.3   Deciding on the techniques to be employed

You will now have to work out the details of the techniques you plan to use for each step of your investigation. The details may be obvious or readily available from, for example, your supervisor, a technician, the literature, a previous project in the Department or from your own knowledge and experience. Alternatively, you may have to work them out or develop new ones which will then be part of your preliminary investigations.

In deciding what techniques to use, your main aim is to ensure that they will work with minimum error and thus enable you to collect results. You will need to try to anticipate possible sources of difficulties, error or problems and adapt your techniques to take account of them. Generally, you are unlikely to be able to anticipate all sources of difficulty and some will emerge as you start the practical work. You can then take steps to adjust the techniques to compensate or overcome them. With the variety of projects available in biology, and in a book of this size, it is not possible to give precise design details of every single step. However, the following notes will prove useful when either making a choice of technique or using it.

FACTORS TO TAKE ACCOUNT OF IN DECIDING ON METHODS

A. *USE WELL-ESTABLISHED, STANDARD TECHNIQUES AND PROCEDURES*—where possible, that are known to work.

B. *TRY TO ENSURE THAT THE TECHNIQUE IS RELIABLE AND EFFICIENT*—that is, that it gives consistent results on the same material and can be carried out quickly (with practice if necessary).

C. *MAKE SURE YOU ARE AWARE OF ANY PRECAUTIONS AND ANY PREPARATORY PROCEDURE*—which might be necessary in using the technique—most techniques have these.

D. *AIM TO ENSURE THAT THE TECHNIQUE CAN BE EASILY REPEATED*—in exactly the same form from practical period to practical period as necessary. This avoids introducing additional sources of error which could affect the material and the results; that is, the technique is consistent between each utilisation.

E. *TRY TO ENSURE THAT THE TECHNIQUES THEMSELVES CAUSE A MINIMUM DISRUPTION OF NORMAL FUNCTION*—of the organism or of a part of the organism. Basically you will be trying to mimic the normal conditions as closely as possible in spite of the fact that it is being prepared in some way; for example, applying physiological saline, which is as similar as possible to the normal body fluids, can help maintain an isolated heart preparation in a normal condition. Keeping organelles cool while being extracted, allowing an animal time to acclimatise before testing it, ensuring the correct nutrient agar is used for culturing colonies, are all attempts at maintaining the normal conditions of the organism.

F. *QUESTION WHETHER THE TECHNIQUE ITSELF WILL ALTER OR MASK A RESPONSE IN SOME WAY*—For example, anaesthetics applied at too high a dose might mask a response; ammonium sulphate precipitation of proteins would alter the measurements of enzyme activity measured in terms of amount of ammonia present; collecting samples from the field at limited times of the day could give false estimates of population characteristics.

G. *ENSURE THE METHOD DOES WHAT YOU REALLY WANT*—That is, when using a method to extract, purify, isolate parts of organisms, show up particular characteristics or make measurements, ensure it really does it. For example, make sure you have the right stain to identify certain histochemical details; ensure the technique really does show up the presence of amino acids you wanted; confirm that the strain of insect you are working on is resistant to the insecticide you are investigating; ensure that oxygen evolution is being measured and not, for example, oxygen utilisation. You may be able to use a technique, such as positive control, to verify that the procedure is doing what it is supposed to do.

H. *AVOID DEPENDING ON COMPLEX APPARATUS OR EQUIPMENT*—If it is not necessary or if there are many difficulties associated with its use, avoid it.

I. *TAKE CARE IN CHOOSING INSTRUMENTS OR APPARATUS*—In deciding what particular instrument to use for any of the steps make sure: it is reliable; you are aware of and can apply the precautions; it operates in the right range for your investigation; it has an appropriate sensitivity and accuracy (note that there is no great merit in having an instrument working well beyond the sensitivity or accuracy needed); it is available when you need it; you understand what calibration and/or standardisation techniques are necessary for using it.

J. *TRY TO ESTIMATE WHAT EXACT EXPERIMENTAL CONDITIONS ARE TO BE APPLIED*—Type; level of treatments; when exactly; for how long; is recovery after the condition has been applied a necessary part of the technique? is more than one factor/condition to be applied and if so how exactly? (take into account the table prepared for recording and analysing results as advised in section 3.4); should the level applied be related to normal levels in the organism? etc.

K. *CHECK ON THE FEASIBILITY OF THE METHODS*—In deciding what methods are to be employed for applying the experimental conditions, ensure that they can be applied to the right level for the right time with minimum disruption (in themselves) to what is being measured/observed.

L. *MAKE SURE THAT THE MEASUREMENT SCALE USED FOR COLLECTING THE RESULTS IS APPROPRIATE FOR THE ANALYSIS PLANNED*—(See section 3.4.) If not, examine the techniques or the analysis and adjust the design as necessary.

M. *AIM TO ENSURE THAT MEASUREMENTS OF THE VARIABLE ARE AS QUANTITATIVE AS POSSIBLE*—If there is not an obvious measurement scale for the variable, consider whether it is appropriate to develop one. Preliminary results will be necessary to help with this.

Example

A student was interested in finding out if different strains of pea plant showed varying susceptibilities to a particular plant virus. He therefore grew the plants and introduced the virus and studied the infection of different parts of the plants. At first, he was going to give written descriptions of the degree of infection backed up with photographs. However, with a little thought he realised that he could devise and use a ranked scale of measurement (see section 3.4) for the degree of infection of various parts of the plant, and used this as well. From his sample of plants he was able to summarise the effect the virus had had and was also able to use statistical tests to test for significant differences in the degree of infection for the different strains. In the end, he was able to make precise conclusions about the relative susceptibility of the strains, which he would not have been able to do so easily or with such precision if he had not used an ordinal scale of measurement.

N. *ALWAYS KEEP FULL RECORDS OF TECHNIQUES PLANNED*

3.4   A consideration of the analysis of the results

An important aspect of design, often missed or underestimated by students, is careful consideration of the methods of analysis of the results. It is worthwhile for two reasons: firstly, it helps to focus your thoughts and ideas on to achieving the aim, and secondly, it ensures that the results are useful, in the sense that they are analysable, as the design takes the analysis into account. You might not be able to plan the *precise* details of the analysis until you have looked at all stages of the design, carried out preliminary work or even completed the full investigation (for example, you may not know whether a variable on a graph should be graphed as a log value). However, aim at the moment to determine the general method of analysis, then come back after you have looked more fully at the design, and clarify the details as far as possible. The following notes will help you in deciding your methods of analysis.

ADVICE ON CHOOSING A SUITABLE METHOD OF ANALYSIS

A. *DRAW UP A TABLE*—which will be used to record the raw data, to show the variable and the conditions of the investigation. It will be clear what can be analysed from it. The table will be in the following form.

Table showing values of the variable for each condition identified

Conditions (*or treatment levels*) of the investigation, such as pH values, species, growth medium, substrate concentration, etc.

		A	B	C	D............
Individuals in	1				
a sample or	2				
replicates (as	3				
necessary)	...				

For your project you may have to include additional conditions (factors) on the table or additional variables which could be recorded on the same table (as long as it is clear which variable is which) or on a separate table. The table is then a record of the raw data from which to make your analyses.

B. *CHOOSE THE MOST APPROPRIATE METHODS FOR ANALYSING YOUR DATA*—by identifying the reason for the investigation, looking at the data in the table and using chapters 5 and 6 which show the general methods used for analysis. Some examples of possible reasons for an investigation with a method of analysis are shown below. You may find it useful and necessary to use more than one method of analysis.

Examples

Identify a trend or pattern of response—graphs
Make comparisons—tables, graphs, charts, statistical significance
    tests
Define characteristics—drawings, photographs, statistical summaries,
    histograms
Determine the relationship between two variables—graphs, statistical
    techniques
Determine significant points or differences—significance tests
Give descriptions—written text, diagrams, photographs, summary
    statistics

C. *MATCH THE STATISTICAL TECHNIQUE TO THE SCALE OF MEASUREMENT USED*—If data are measured on a nominal or ordinal scale of measurement, non-parametric techniques should be applied, whereas for data with a ratio or interval scale of measurement, parametric and non-parametric techniques can be used, and preferably the former. The measurement scales are given in table 3.1 and it is worth identifying the one you plan to use so that you can choose the correct statistical technique.

Table 3.1
Scales of measurement

(i) NOMINAL OR CLASSIFICATORY—Objects or individuals are classified into mutually exclusive and totally inclusive classes; for example, they could be classified on eye colour (brown, green or blue), or on whether they germinated or not. The usual form of the data in an investigation is that numbers, counts or frequencies having a particular characteristic are grouped into a particular class; for example, in a sample of 50, 20 individuals have blue eyes, 19 brown and 11 green. Individuals can never be placed in more than one class and the boundaries of the class are clear cut enough for it to be clear to which class each belongs.

(ii) ORDINAL OR RANKING—Individuals are classified into mutually exclusive and totally inclusive classes but here the classes can be arranged in some order so that they are in some relationship to one another; for example, the classes may be arranged to be 'greater than', 'stronger than', 'better than' etc. It is useful for further analyses to give the classes a number reflecting an order so, for example, the degree of infection might be measured on an ordinal scale from 1 to 5, where 1 is no infection, 2 more infection, 3 even more infection and so on. It is important to be clear

about and to recognise the characteristics of each of the classes so that individuals can be assigned to only one class and consistency maintained whenever the scale is used. In using the scale of measurement in an investigation, it is common to use a sample of individuals so that each individual is given a value (class) from the ordinal scale; hence, in surveying a group of infected animals one animal may be given the value 1, another 5, another 3, etc.

Ordinal scales of measurement are particularly useful in animal behaviour where they can be used to measure behaviours like degree of aggression, motivation, activity, etc. They are also useful when you need to quantify characteristics like healthiness, amount of colour, degree of damage or inhibition, amount of growth, etc., when other measurements are not possible or practical.

(iii) INTERVAL AND RATIO SCALE—These are scales which have all the characteristics of an ordinal scale but in addition the distance or interval between any two numbers on the scale is of known size. Underlying the scales is the assumption of 'additivity'. This assumption is that the distances between observed points can be added together to give the total distance; that is, they can be added and subtracted. They are truly quantitative scales and defined by units of measurement. When the scale has a true zero point as its origin, it is called a 'ratio scale'; if not it is called an 'interval scale'. For example, scales measuring weight are ratio scales and a scale measuring temperature on the Fahrenheit scale is an interval scale. In both instances, the interval between any two numbers is known; for example, the difference between 6 g and 10 g is 4 g; between 10°F and 25°F is 15°F. (This contrasts with an ordinal scale of measurement where the difference between 1 and 4 on a scale of aggression is not 3 aggressions.) In the former instance, 0 g is 'no' weight so the scale has a true zero and is a ratio scale. In the latter 0°F is not 'no' temperature so the scale does not have a true zero at its origin and is designated an interval scale.

D. *CHECK WHETHER THE METHODS OF ANALYSES IMPOSE CERTAIN CONDITIONS ON THE DESIGN*—For example, if a Chi square contingency table is to be used, minimum frequencies (numbers of individuals) are necessary for each category/class. If a trend or pattern of response is wanted, a good range of treatments will be required (the number of samples needed should be obvious from the table prepared for the results as should the type and size). The following examples give some brief details of how the aim determines the analysis, which in turn makes certain demands on the design.

Example

In all the examples given below the yield of sprout, plants, measured as total weight of the sprouts after a set period of growth, was the variable investigated. The plants were subjected to fertiliser brands which were being developed and the overall aim was to assess the fertiliser's efficiency.

aim (i)   *To test a range of concentrations of fertiliser on the yield*

ANALYSIS: Graph of yield versus concentration of fertiliser will show the trend.

DESIGN CONSIDERATIONS: A good range of fertilisers wanted, so 10 concentrations chosen from 1/4 to ×4 the recommended dose. A sample was investigated for each concentration and the mean yield±standard error bars were to be placed on the graph. (Samples used because of the problems which would occur in growing plants to a particular stage of development.)

aim (ii)   *To test whether the recommended dose of fertiliser increased the yield*

ANALYSIS: Statistical significance test for making a comparison to determine whether there are significant effects. 't' test chosen since measurement scale is appropriate and two samples compared.

DESIGN CONSIDERATIONS: Two independent samples needed, one a control and one receiving the fertiliser, as similar as possible in all other respects.

aim (iii)   *To compare the yield of plants subjected to two different brands of fertiliser to determine whether there were differences between the fertilisers*

ANALYSIS: Statistical significance tests for making comparisons between groups to test for significant differences. One-way analysis of variance chosen followed with multiple range testing.

DESIGN CONSIDERATIONS: It would have been possible to use two samples as before, each receiving a different fertiliser, and testing for significant differences with the 't' test. However, since the project is overall concerned with an increase in yield which had not been established, it was thought useful to include a control. Three samples therefore needed: one a control, each of the others receiving a different fertiliser, as similar as possible in all other respects.

aim (iv)   *To test the effect of three commercial brands of a fertiliser on the yield of four varieties of sprout; to compare the yields between the varieties and the brands*

ANALYSIS: Statistical significance test for comparisons for significant differences. Two factors involved, more than two samples, therefore a two-way analysis of variance chosen.

DESIGN CONSIDERATIONS: Consideration of the way the results need to be recorded indicates that nine samples will be needed (see below), each subjected to different treatments (SIA, SIIC, etc.):

		\multicolumn{3}{c}{*Brand of fertiliser*}		
		A	*B*	*C*
Variety of sprout	I	SIA	SIB	SIC
	II	SIIA	SIIB	SIIC
	III	SIIA	SIIIB	SIIIC
	IV	SIVA	SIVB	SIVC

Replication for each of the conditions is not absolutely necessary but thought to be useful here as tests for interaction between the two factors could therefore be investigated. The sample sizes need to be the same for this significance test. If a value is lost (that is, a plant is destroyed for some reason) there are techniques for producing replacement results. Control not included in this instance as all fertilisers had been shown to increase the yield of these varieties and the overall aim is to compare the varieties to see which is best for the brands. Recommended doses (field doses) were used.

3.5   Controls, samples and replication

At this stage you should have a fairly clear idea of how you plan to analyse your results, at least in the broad sense, the steps you need to perform and the techniques you will be employing in each of the steps. There are two main aspects of design which need further consideration: controls and samples.

3.5.1   Controls

Controls are generally required in an investigation which is a 'true' experiment in which the experimenter alters the level of some factor (the experimental treatment) in a controlled way. The aim of the experiment is to determine what effect that treatment has. The effect can be determined by comparing this experimental situation with a control which is identical in every possible way, except that the specific experimental treatment is not applied. Any differences between the two can then ideally be attributed solely to the experimental treatment.

Example

If the effects of a drug were to be tested, the method may be to dissolve the drug in saline and then inject into a particular muscle. The control would be to inject at the same time an equal volume of saline into the same muscle of a different animal, pre-treated identically and selected from the same population, in the same way. Differences between the two animals would therefore be minimised as far as possible so that any differences between them are likely to have been caused by the presence of the drug in the experimental animal.

In certain experiments there is no obvious control as described above. This happens where the factor being altered in the experiment cannot sensibly be eliminated or is essential for normal function. For instance, if the experimental factor was temperature, it is implied from the above that a control would have no temperature—that is, 0° on the Kelvin scale. This makes little sense. In such situations it is usual to regard the normal conditions of the organism as the control or to define the aim differently. For example, rather than 'testing' the effect of temperature the aim may be 'to compare the effect of two temperatures' or the control could be the ambient temperature. An expansion of this kind of experiment would be to test a range of levels of the experimental factor (such as a range of temperature or pH), the aim being to determine how the organism responds. Once again, a control, in the sense originally indicated, would not be applicable. However, in each instance the same kinds of principles apply—that is, the conditions should be as identical as possible for each level of treatment applied, with the only difference between them being the level of the treatment.

In setting up a control, aim to minimise differences between the experimental and the control. Take a long, hard look at the two conditions and ask yourself how they are different. If the only differences are the treatments to be tested, it is OK. If there are other differences, then you need to redesign the control to overcome or take account of them. Some differences which can easily be missed and overcome, once recognised, include such things as: timing; general conditions during the experiment, such as the position on the bench; type of containers; batch (population) of organism; and preparatory procedures.

In an attempt to minimise difference between the experimental and control conditions the same organism may be used in both situations, thus receiving both conditions. Great care must be taken to ensure that neither the testing procedures nor the giving of a response alters the state of the organism for retesting.

Example

The effect of sex hormones may be tested on sexual activity measured as the number of successful matings by a male mouse in a set period of time with 20 females. As a control the number of matings without the hormone could be measured, then the hormone injected and the number recorded. However, the male may have learned from the previous experience which increased his chances of being successful in the next instance.

Some precautions which can be taken include: altering the order in which the two conditions are tested (crossover experiments); allowing time for recovery to see if the control response is once again recorded; and in certain kinds of experiments, using standard procedures to cope with many of the common problems (for example, pharmacology dose response experiments). Alternatively, the problem may be avoided by using matched but separate individuals (see page 47). Samples of individuals rather than one individual are usually investigated in an experiment—there being a control sample and an experimental sample of the same or similar size.

In some instances, it is useful to have a positive control. This is a control which aims essentially to confirm that a particular technique is working in the way it should. For example, if using ninhydrin to determine the presence of amino acids in an extract, the positive control would be to add ninhydrin to a solution containing known amino acids to confirm that the ninhydrin does show their presence. It is particularly important in instances where negative results are unexpectedly obtained.

3.5.2   Samples

A common question facing investigators is 'Should a sample be used?' Invariably the answer is 'Yes', though in certain instances it is less appropriate and examples of such cases are dealt with in section 3.5.4. The reasons for using samples are outlined below. (Accounts given in the statistics books cited at the end of chapter 6 are also very useful.)

In order to understand the biology of organisms we, as biologists, are interested in being able to make some general statements about the characteristics of identifiable groups, defined as populations. So, for example, we might wish to define the response characteristics of vertebrate striated muscle to stretch, we might want to identify the typical structure of a species of insect, define the growth characteristics of a bacterium, or describe certain chemical reactions occurring in liver cells of mice. We are less interested in *one* individual—for example, how one particular nerve conducts impulses, or how a single, particular geranium plant responded to a lack of nutrients. However, an outstanding feature of biological organisms is their variability so that fine details of characteristics do vary from one organism to another. The major difficulty facing biologists, therefore, is to develop procedures which take account of this natural variability between organisms in order to be able to make some general statements in spite of it. The main way we can do this is to take samples of individuals, summarise their characteristics, and use these summaries as being representative of the population as a whole. In order for such an assumption to be reasonable, samples must be taken from the population with care so that they reflect as closely as possible the population characteristics. Such samples are random samples in which each individual has an equal and independent chance of being selected. Other properties of the sample which need to be considered for an investigation are sample size and sample type. Armed with the information from the samples, statistical techniques are then employed to summarise the data and to use these to make inferences (generalisations) about the population as a whole. Comparisons can then be made between populations.

Given below are a list of procedures you would find useful to employ in selecting a sample.

PROCEDURES TO EMPLOY WHEN SAMPLING

A. *DEFINE YOUR POPULATION CLEARLY*—You need to be clear about the nature of the population you are investigating, since: it is its characteristics you are aiming to define; you may wish to sample from the same population again; or others may wish to repeat or carry on from your work. Basically, you need (as a minimum) to give the species of organism, the strain, breed, variety, etc. If you have been more selective, identify the features you have used for selection, such as weight, sex, viability, and size. In addition, be aware of your source and the conditions for maintaining the organism; for example, organisms were obtained from supplier *x*, kept in individual cages where food y^2 and drink *ad libitum* were supplied till they reached a weight between 350 g and 400 g.

B. *USE AN OBJECTIVE METHOD FOR OBTAINING A RANDOM SAMPLE*—The basic procedure is to assign a series of numbers to individuals in a population and then use random number tables to determine which individuals will be included in your sample. The technique is to start at one point on the table, work systematically in one direction and record the numbers till you have enough for the sample size required. Then select those individuals which have those numbers. In ecology there are often well-documented procedures for sampling from populations based on these principles and you should obtain advice from your supervisor if yours is an ecology project. When assigning treatments (such as control or experimental) to individuals, a random procedure should be used. Sometimes it is not possible or practical to obtain truly random samples—for example, if you need ten mice and there are only ten in the animal house, you will have to use all of them; if you have a population of single cell algae in culture and want a sample, you cannot number every individual. In such cases try to apply the criterion of a random sample as well as possible—that each individual has an equal and independent opportunity of being selected; therefore, in the alga example give the culture a good stir or shaking and sample from the mixed group, and do not pick on the big ones or little ones, or ones from particular locations. A worthwhile alternative, if obtaining random samples is difficult, is to take another sample drawn from the same population and check whether similar results are obtained. This second replication of the experiment may compensate for the inability to draw a random sample and assumes that it is unlikely that the same biases and oddities as the first sample are present. If the same results are obtained then there can be reasonable confidence that the results are the 'real' thing. Similarly, if your results are confirmed by, or a confirmation of others' work, there can be confidence that they are 'real' findings and not due to the biases of sampling.

C. *DETERMINE WHAT SIZED SAMPLE IS NEEDED*—This is often difficult since what you really want is a sample size which is adequate for identifying population characteristics, so it should be neither too small to mask those characteristics nor too large to waste effort and time. The size depends to a large extent on what your investigation is attempting to show and on the analyses used; for instance, if using Chi square tests, generally minimum frequencies are necessary in each of the classification categories for the test to be valid; if using significance tests, there are procedures for determining the sample size needed to obtain significant differences but these largely depend on having identified the population characteristics in the first place.

If you are planning to use statistical techniques it is worthwhile appreciating that:

(i) statistical techniques do take account of sample size in conclusions of significant difference and in estimating the standard error (the outcome is that with small samples there will have to be bigger differences between groups for a significant difference to occur than with larger sized samples);

(ii) most statistical tests are robust in the face of violations of assumptions if large samples are involved (or more specifically if based on a large number of error degrees of freedom);

(iii) as the sample size gets larger, it has increasingly smaller effects on the standard error (a measure of the precision of the estimate of the mean) since the standard error decreases in proportion to the square root of sample size.

SOME OVERALL GUIDES FOR DETERMINING SAMPLE SIZE

(i) AIM FOR SAMPLE SIZES OF BETWEEN TEN AND THIRTY—Try not to go below five. If you have a choice, complete an investigation with a reasonable sample size rather than doing another investigation; hence, rather than doing two investigations with five in each sample do one with ten.

(ii) KEEP THE SAMPLE SIZES THE SAME—if you are comparing groups. So, for example, make sure that if you have a control and experimental group(s) the size of the sample is the same for each.

(iii) MATCH THE SAMPLE SIZE TO THE PROBLEM IN THE INVESTIGATION—For example, ecology experiments concerned with defining population characteristics in which a histogram was to be drawn to show the frequency distributions of the variables may require large samples.

(iv) SEEK ADVICE FROM YOUR SUPERVISOR OR STATISTICIAN—if in doubt.

(v) REPEAT THE INVESTIGATION WITH THE AIM OF INCREASING THE SAMPLE SIZE—if on completing the significance test, probability values are close to significance but not significant, for example, 10 per cent (however, you may have to confirm that the second sample is giving the same results as the first before combining data). You should then be able to show which conclusion, significance or non-significance, is the most acceptable from the data.

D. *DECIDE HOW MANY SAMPLES ARE TO BE USED AND OF WHAT TYPE*—If you are comparing two or more groups you will need a sample for each group. The samples across the groups can be independent (unmatched samples) or matched samples. In an independent or unmatched sample, each individual in the sample is unrelated or independent of any individual in another sample. In a matched sample, an individual in one sample is matched as closely as possible, and therefore related, to an individual in another sample; that is, the first individual in a sample is matched as closely as possible to the first individual in the other samples, the second is matched to the second in the others, the third to the third in the others, etc. Significant tests on matched samples are concerned with examining differences between the individuals in the matched set, whereas significance tests on independent samples are concerned with testing whether the difference between the groups is greater than the differences within each of the groups. The matched design therefore excludes a major source of variation in biological material from the comparison, that is, that between individuals in a random sample, and so considerably enhances its sensitivity. Consider whether a matched design would be useful for your investigation. Matching individuals across the groups will vary according to your experiment, so it has to be done carefully. An obvious matching device is to use individuals from the same litter, or eggs from the same female, or seeds from the same flower; another is to use the same individual (as long as it is confirmed that the treatment given in one of the groups does not affect the response in the remaining treatment groups). Alternatively, it might be more appropriate, in your experiment, to achieve the matching based on size, age, source, previous history or experience. Examine carefully what you are aiming to do in the investigation and do the matching appropriately. Though individuals within a matched set match as closely as possible, the different matched sets can vary widely. Once a matched set has been fixed, treatments are allotted to each individual in the matched set at random.

3.5.3   Replication

The techniques so far described in this section on samples have been concerned with showing the value of taking samples and how to obtain them. Such sampling is primarily concerned with coming to terms with the variation between individual organisms. However, in biological investigations there can be another serious source of error related to the procedures themselves—for example, an instrument may show serious instability, so readings can vary. Aliquots taken from the same material extracted or collected from an organism may show variations. In such instances, replications are useful for coping with the possible sources of error where a replicate is a repeat of the same procedure, and in this instance a repeat on the same material from the same organism. Various statistical techniques, usually a mean and standard error, can then give the average value.

Example

Suppose you wished to work out the K_m of alkaline phosphates extracted from pig liver. Liver from one pig could be obtained and the enzyme extracted and the K_m determined. A replicate would be if the K_m was then confirmed with an extract from the same liver, and would basically be a confirmation that the procedures were working adequately. A sample would begin to be made up if a second liver from another subject was then examined.

> *Example*
>
> Suppose you were trying to test whether there was a cycle of Na^+ excretion in humans. Urine samples could be collected from one individual and an estimate of Na^+ made. To be able to state the level with some confidence, replicate readings of Na^+ levels would be made by taking additional aliquots from the same urine sample. A sample of individuals would be obtained if more than one person's urine was collected and examined.

3.5.4   Special cases

There are certain instances where increasing the sample size or replication, in the sense described, are not the best techniques for coming to terms with the variation. This is particularly the case where graphs are drawn which are standard curves in some form or another, such as a standard curve for a chemical estimation, a growth curve, a dose response curve, curves used in enzyme kinetics, or a strength duration curve of a nerve action potential. In such instances the curves normally fit a well-defined mathematical description and are used for making some sort of estimate of a particular characteristic — for example, optimum pH, threshold response of a nerve, or K_m. The aim of such investigations is to ensure that the best curve can be drawn through the points so that the various estimates can be made reliably and with precision. In these kinds of instances it is a better strategy to increase the number of points on the x axis (that is, the independent variable fixed by the investigator and considered to be the experimental treatments) than to increase the number of replicates or sample size for each treatment. There is another instance where replication is not necessary and this is in investigation involving two factors in which the analysis to be used is Two-Way Analysis of Variance without replication or the Friedman test, in which the primary concern of the investigator is to determine whether there are significant differences between the treatments (Factor one). It is not a suitable design for testing for significant differences between the matched sets (Factor two) or for testing for interaction. If this is required, replication is necessary for each of the combinations of the two factors and the suitable analysis would be a Two-Way Analysis of Variance *with* replication.

3.6 The final design

Having completed these stages of the design you should be able to draw up tables for recording your raw data which show the variable(s) to be observed (units, too, if possible for measurements), each condition of the experiment (including the experimental and control conditions, or a description of the population investigated), spaces to record the data for samples and/or replicate readings as necessary. It should then be obvious how many samples are required and whether they should be matched or not. You should also have a list of the steps necessary to carry out the investigation, details of the techniques to be used at each step and a statement of what kinds of analyses will be carried out to satisfy the aim of the investigation. If parallel investigations are planned, the design information should be available for each parallel investigation.

With the design ready, practical work can start and advice is given in the next chapter.

4

The Practical Work

This chapter will help you to:

Plan your practical work efficiently

Decide what preliminary investigations should be carried out

Organise your practical time for maximum benefit

Carry out the practical work effectively

Overcome practical difficulties

4.1   Some general considerations

Having completed the design to a reasonable degree you should be able to start the practical work. This needs to be carried out in a planned and organised way to maximise the time and facilities available to you and thus to increase your chances of obtaining results. It is, of course, possible to work in a haphazard manner with little forethought and direction but this will be wasteful of time and facilities and probably very frustrating. During the practical work you will become increasingly familiar with the procedures and make various adjustments and refinements. You will also increase your competence at the techniques, general understanding of the project and what is involved in obtaining results. It is important therefore to start the practical work as soon as possible and not spend a disproportionate amount of the timetabled practical time reading the literature or designing the 'perfect' investigation. *Do* recognise that you will have to work in stages, in a systematic way and that you cannot expect everything to work initially, continuously, or that you can complete an investigation within one practical period.

In planning and organising a practical investigation you will have to think through carefully exactly what you need at each of the stages to put the design into practice, and then make sure: that you have the material as necessary in the quantities you require at the time you want it; that you are confident in the use of any specialised apparatus, techniques or instruments; that you have the facilities to store material or set up long-term investigations as needed; and that you can complete any work in the time available.

When planning and organising practical work it is useful to break it up in the following way.

OVERALL APPROACH TO THE PRACTICAL WORK

A. *DECIDE WHETHER PRELIMINARY INVESTIGATIONS ARE FEASIBLE, NECESSARY OR USEFUL*

B. *ORGANISE A SCHEME OF WORK TO CARRY OUT THE PRELIMINARY AND FULL INVESTIGATIONS*

Section 4.2 looks in more detail at various aspects of preliminary investigations, and the remaining sections are concerned with advising you on how to carry out practical work efficiently to derive maximum benefit from it.

4.2   Preliminary investigations

Preliminary investigations are basically test-runs of the design in which you, as the investigator, go through all the steps of the procedure and make sure that each works—that is, that each does what is expected or required; for example, you may need to check whether the sampling procedure produces an unbiased sample, you may wish to confirm that a technique can be applied to the organism you are working with, etc. In almost all investigations unforeseen difficulties and unanticipated problems do arise, so the preliminary investigations are an important first stage of the practical investigations to develop the design to a sufficient degree to ensure that the procedures will work, and work in a sound, scientific way.

Example

A student was interested in comparing the effectiveness of Fenitrothion, an insecticide, on two strains of grain weevil—one resistant and one susceptible to the insecticide. The plan was to place the weevils on filter paper which had been soaked in Fenitrothion and record the numbers dead over time. Three major problems arose; animals congregated in one area of the petridish and did not move (though the experimenter wanted them to move); it was difficult to determine when the animal was dead; and the moistness of the filter paper unexpectedly seemed to affect the effectiveness of the Fenitrothion in killing the insects. The experiment had to be redesigned to take such difficulties into account.

The main value of the preliminary investigations, then, is that they can highlight difficulties and provide possible leads about how to overcome them. In general they are of major importance when procedures are not well tried or established for the organism you are working with—the chances of unforeseen difficulties or problems arising are greater. For your project, therefore, identify what is a new procedure and try to check that it does work.

However, it might not be possible to check procedures through in separate preliminary investigations because of the time involved.

Example

In a project lasting 15 weeks, plants may have to be grown under defined conditions for 8 weeks or animals undernourished to certain stages of development, 12 weeks for instance. There is no time to do trial runs as preliminary investigations.

If you find in your project that you are unable to perform

preliminary investigations, the best thing to do is to make sure the design is as good as you can make it, and that you have made extra special efforts to try to anticipate difficulties and sources of error. Try to take account of them in the design, and start the practical work as soon as possible. You will then have to cope with the difficulties as they arise and keep full records of what happens at each step. In such cases the distinction between preliminary investigation and the main investigation is not therefore very clear cut. What might happen is that you go through the procedures and 'learn as you go on', thus developing and expanding on the techniques. Invariably the first practical work carried out will be less satisfactory than the final work. (You will have become more skilled in handling the material and confident in what you were doing as the project develops.) However, you will often find that you need the results of the early work to present in the report, even though it is not completed as vigorously as you would have liked. The best thing to do is to try to get as much as you can from the final investigations and then use the first set of investigations to substantiate these final results, though recognising the weaknesses.

Example

A student planned to test the effect of metaldehyde on the oesophagus of *Arion ater*, the garden slug. Initially it was difficult to get spontaneous activity to test the effect of metaldehyde. When it was obtained, metaldehyde had such a marked effect that the gut did not recover. Other doses were tried and it took time before a working dose could be used. It also took time before the technique was perfected to give reliable and consistent spontaneous activity and then on a different part of the gut. Towards the end of the project time the student had the whole 'set-up' working well but had little time to get a 'good' set of results with adequate controls and a range of doses. But it was possible to establish a consistent response to the metaldehyde. This pattern was evident but not so clear cut in earlier recordings. The student recognised the weakness of earlier work but needed the results to present in his report as he did not have time to complete the project as well as he would have liked. In reality, the preliminary investigations were all those involved in getting the technique to work in a consistent way and then the main well-controlled experiments began. In his report he gave an account of the techniques he developed and the results obtained, showing the strengths and weaknesses and what had been done. He obtained a good mark.

In contrast to this, there are a number of instances where preliminary investigations are particularly useful for clarifying some specific aspect of the technique. In such cases the preliminary investigations are often very clear cut and separate from the main investigations. Some examples of the value of preliminary investigations in such instances are given below.

SPECIFIC USES OF PRELIMINARY INVESTIGATIONS

A. *CONFIRMING DETAILS OF THE TREATMENTS OR CONDITIONS*—Preliminary investigations may be necessary for determining the precise levels of treatments to apply, for example, in a particular toxicology experiment it was not clear what dose of toxic substance should be applied and for how long. The preliminary experiments could be used to establish these parameters for the investigation. Alternatively, it might be necessary to establish ideal conditions for storage, extraction and maintaining isolated tissue, etc.

B. *DEVELOPING A TECHNIQUE*—The preliminary investigations may be necessary for developing or adapting a technique for any of the steps in the procedure. For example, it might be to develop the best extraction methods for a certain compound, or the best technique to get a stable recording, or the best method of infecting an organism, or separating some proteins, or finding a method to test choice by animals, etc.

C. *ESTABLISHING A MEASUREMENT SCALE*—In the design you may have planned to quantify some parameter using, for example, an ordinal scale of measurement, but need either the organism's response or for it to be present to establish what each point on the ordinal scale represents; for example you may wish to establish an ordinal scale of measurement of degree of sliming by a snail in response to air currents. Preliminary investigations could be designed to establish such a scale. In devising an ordinal scale of measurement, it is extremely important to make sure that each category or value in the scale is clearly defined so that the scale can be reliably applied over a period of time—that is, make sure you are clear how a value of 1 on the scale differs from 2, and that a value given 1 at the time is also given 1 at another time. One way to standardise is to have clear descriptions or a set of standards for each value in the scale and to test whether on different occasions the same score is given. In devising nominal scales of measurement, make sure again that the categories are mutually exclusive and totally inclusive.

D. *PRACTISING THE TECHNIQUES*—You may need practice at the procedures from sampling through to recording; for example, you might wish to confirm that you can withdraw a certain volume of body fluid from a crab without severely damaging the animal and that you have enough to carry out an analysis. In preliminary experiments you could practice withdrawing the fluid without subjecting the animal to the experimental conditions, or you might wish to confirm that you could identify different groups of microorganisms before examining populations under experimental conditions.

E. *BECOMING FAMILIAR WITH INSTRUMENTS*—It is always useful to spend a period of time making sure that you can operate the instruments or apparatus fully and correctly. It is worthwhile therefore obtaining the instruction manual, going through it and making sure you know what the various switches are for. Be clear what standardisation or calibration procedures are required to operate the instrument and do some practice runs on similar material to that needed for your investigation; for example, collect some nutrient solution to measure the Na^+ level on a flame photometer, or collect some leaf material to measure the energy level in a bomb calorimeter. Be aware of the precautions necessary in operating the machine and make sure you can put them into practice.

F. *CHECKING PROCEDURES*—You may wish to confirm that a particular step in the procedure does what it is supposed to do; for example, you might wish to check whether the technique used to obtain a sample really does give an unbiased sample—this is especially important in ecological investigations. You might wish to make sure that the activity monitor really is able to monitor the activity of the cockroaches over a period of time, or that you can set up different wavelengths of light and yet keep the temperature constant.

If you are using preliminary investigations for these kinds of reasons, set them up as soon as possible and work specifically on them without setting up the experimental procedures in detail. For example, if you plan to extract an enzyme from a biological source and measure its activity, you might want to confirm that the method for assaying the enzyme works in the preliminary investigations. Instead of using the extracted enzyme which has taken time and effort to obtain, use a commercial preparation of the enzyme to establish that the assay works first.

The difficulties encountered in preliminary investigations might be relatively easy to overcome, like for example adjusting a pH or a temperature, or they could be extremely difficult with the result that the investigation may be modified substantially and perhaps take a different line. It is quite reasonable and common for projects to take modified lines of investigation from that originally proposed, as a result of the practical work. The modification might be relatively minor as, for example, adjusting the levels of an experimental condition slightly, from 20°C to 25°C, or could be quite drastic as the following example shows.

Example

A student was interested in examining biochemical responses to pollutants in a marine alga. The plan was to make extracts of the cellular fluids to determine the levels of various biologically important compounds—carbohydrates, nucleic acids, etc., and eventually to test the effects of various pollutants on these levels. Practical work began but it proved very difficult to break open the cell walls and release the cell contents. The project then took a different direction and became a study to determine the most effective method of releasing cell contents from the marine alga cells.

The newer the procedures for a particular organism, the greater the chance that the investigation will take a different line of enquiry.

All in all, preliminary investigations are a very useful and valid part of a project which you may be able to carry out satisfactorily if you have enough time. Very often the results of preliminary investigations can be justifiably presented in the final project report as an essential component of the project.

4.3 Planning and organisation

Listed below are a number of points worth considering for planning and organising the practical work in your investigations. These points cover the common problems encountered and mistakes made by students, and take account of the kind of information students themselves commonly expressed a need for. You should find them helpful and useful.

PLANNING AND ORGANISING THE PRACTICAL WORK

A. *ASCERTAIN THE ROLE OF THE TECHNICIANS AND SUPERVISOR*—It is probably true to say that for your course as a whole the lecturers and technicians have taken on the responsibility for planning and organising the practical work in detail. It has therefore been their responsibility to ensure that you know what you are supposed to be doing in each practical period, that the material is ready for use, that instruments are working and that the work can be completed in the time. In the project all these responsibilities are now *yours. You* will have to make arrangements to ensure that any material you require is ordered to arrive on time, *you* will have to make sure that the apparatus or instruments you need are available and working, *you* will have to make sure you can complete the work you set yourself in the time you have got, and *you* will have to make up your own solutions. If you are not aware of these things it can be very disconcerting, for example, to ask for a 0.9 per cent solution of NaCl and be given a jar of solid NaCl to make up yourself—especially if you do not know how to do it. Check, therefore, very early on what exactly you can expect from the technicians. It is probably best to assume that they will do little preparation for you and you will have to do just about everything yourself. You can therefore plan your practical work with this in mind.

B. *MAKE LISTS OF THE ITEMS YOU NEED*—Make lists which show exactly what you will need for each step of the investigation. One way to do this is to imagine yourself doing the practical work in stages and think what is needed; for example, consider such things as test tubes, filter paper, distilled water, slides, cover slips, chromatography tank, stop clock, Ringers solution, spectrophotometers, etc. For each item identify what quantity is necessary each time you carry out the step. Sort out what is routinely available in the labs, what you will have to ask for (because, for example, it is kept locked away) and what you will have to order from outside the department. Distinguishing between these last two may be a bit difficult but a check with the technician should sort it out for you. If there are solutions you need, check if they are standard solutions, routinely available, or whether you will have to make them up yourself. If you have to make them up yourself, you will then need to list the necessary chemicals.

C. *MAKE ARRANGEMENTS TO ORDER MATERIAL*—This is particularly important if the material has to be ordered from outside the Institution as there could be some delay in having it delivered. In placing such orders, consider what quantity you will need for the duration of the project (not just one practical session) and order enough to see you through. You may need a regular supply, such as some fresh organisms once a week, so place the order to ensure that you do have a regular supply. If you have to place orders in advance for material held in the department, such as chemicals or apparatus, do so.

D. *BOOK SPECIALIST EQUIPMENT*—In your department there may be certain procedures for using some of the specialist equipment; for instance, there may be a booking system or it may be necessary that a member of staff or a technician is present while you are using it. Check, therefore, what those procedures are and plan your work with those in mind. If there are no particular procedures, confirm that the instruments will be available for use when you require them and that you have the operating instructions (booklet if possible) in case anything goes wrong or for using them to maximum advantage.

E. *PREPARE AS MUCH AS POSSIBLE IN ADVANCE*—For each practical session, preparations can be made which help save time or help you to use your time more efficiently. For example, in many projects it is possible to make up material in bulk so that it is readily available in each practical period and does not have to be made up each time; examples are standard chemicals, agar, nutrient solutions, etc. Look through your lists and see what you will need routinely. Check whether the material can be safely stored and if so make it up in bulk in quantities you will be needing through the project and make arrangements to store it in the appropriate conditions. If material or organisms have to be prepared in some way ahead of time, make early plans to do this; for instance, you may have to acclimatise invertebrates to certain temperatures.

F. *SET UP APPARATUS BEFOREHAND*—If possible in your department, try to set up all the equipment you need before you start the practical work. When doing this make sure everything is working; for example, if you wish to record from nerve cells you could connect together all the various instruments and ensure they are working. At the practical all that need be done is to collect the animal and then to record from the nerve cells.

G. *CHECK WHERE IT IS SAFE TO HAVE A BREAK IN THE PROCEDURE*—You may not be able to carry out all the steps of your procedure during a practical session. Sort out in advance, therefore, where it is safe to stop and store the material and also determine what conditions are needed for storage. Remember then, when organising your practical work in the practical sessions, you will have to get to these stages where material can be stored.

H. *ORGANISE STORAGE FACILITIES*—It may be necessary to set up long-term investigations or to store material. Certain conditions may be necessary for these. Make sure that the facilities are available and suitable for your investigation; for example, you may have to keep animals on a reversed photoperiod, or grow microorganisms at fixed temperatures, or store material in a freezer within a set temperature range.

I. *PLAN A SCHEME OF WORK*—For each practical session have a clear idea of what you aim to achieve in it and know where you plan to go next. When you plan each session, be realistic about the time that might be needed to complete the work and aim to do enough to fill the time, but not too much so that you cannot complete it or have to do it in a hurried way (mistakes are likely to creep in). It will not be possible to plan every practical period in advance, but you should always aim to plan ahead and thus have an idea of what you plan to do in two or three consecutive sessions; for example, in the first session you might decide to familiarise yourself with some apparatus, in the next to prepare all the material you need, in the next to run through all the procedures to check they work. If you therefore finish some work quicker than expected you may be able to start on the next stage in an organised way. If your organisms require a period of pre-treatment or if there is any other long preparatory period, make sure that this is started early on; for instance, you may have to grow plants for 6 weeks, you may have to culture organisms for a month, etc. One advantage of planning a scheme of work is that you can progress and see your progression, which can be very reassuring, particularly if there are a number of procedures to carry out before results are obtained.

J. *BE AWARE OF THE SAFETY REGULATIONS*—There will be safety regulations available for working in your labs which you should familiarise yourself with.

4.4   Doing the practical work

The following points are aimed to give you maximum benefit from the practicals so that you can get results. They take account of many of the mistakes or poor codes of practice I have noticed in students' work.

GENERAL ADVICE ON DOING THE PRACTICAL WORK

A. *KEEP A PRACTICAL BOOK*—There is a tremendous temptation to record work of the project on scraps of paper. These get lost and are often so disorganised that it is not clear what exact information is being recorded. It is extremely annoying and frustrating to lose, mislay or not be able to decipher vital bits of information, and research workers very quickly get into the habit of being disciplined in recording their work. The best way is to have a practical book in which you record information about the project. Its main use is to record results of your practical investigations. Routinely you should note the date and time of the record since these may prove to be significant in the interpretation of the results. You should also make sure that the records show clearly what exactly is being recorded (variable), the values of those measurements and the units (often forgotten). Include, too, any other information which may be relevant, such as weight of organism. In addition, the book can record other useful information about the project: perhaps ideas you have about the interpretation of the data being recorded; or about further useful investigations which could be carried out; or some general comments relating to the recording of the results, like the difficulties encountered or unusual responses. I find it useful to write a summary of the day's work while it is fresh on my mind for future reference. This book can ultimately be quite a valuable source of ideas and information when interpreting results or planning future investigations, and for giving an account of the methods.

B. *RECORD CAREFULLY AND HONESTLY*—Be alert while collecting results and make sure that the records being made are accurate and correct; for instance, if the temperature in the water bath is 21°C though it should be 20°C, record that it is 21°C. If there are difficulties in making measurements or if you are uncertain of a result then make a clear note of it; for example, if you had difficulties in identifying whether the seed had germinated or not, or was viable, then note it.

C. *AIM TO WORK ACCORDING TO THE PLAN AND THE DESIGN*—You have put a lot of thought into the design and plan of the practical work so try to work to that plan and not be easily diverted from it—for example, by a strange or unexpected response. If that does occur take a note of it, find out what you can without jeopardising your investigation, and plan to consider it further (if you would like to) with another investigation which is designed to do so. If, however, your design is completely inappropriate then make sensible adjustments; for instance, if you needed to look at the gill beat characteristics of a freshwater crustacean and the conditions used stopped the beat completely, then you would have to adjust the conditions so that you could make measurement and these may not be the conditions you planned for, or if you were examining aspects of feeding behaviour and the animal would not feed you would have to reject that animal and use another so that you could investigate the characteristics. Records of what happened and adjustments made would be kept.

D. *COLLECT TOGETHER EVERYTHING YOU NEED*—Try to ensure that you have collected together all the material necessary to carry out the planned practical work so that you can avoid going off to search for equipment or material at what might be a crucial point in the recording. Once you have collected everything together, then begin the practical.

E. *CALIBRATE EQUIPMENT*—Remember to calibrate equipment and to take note of the scales being used to record data. It seems quite common to forget to calibrate for tension in experiments involving muscle for example, or to calibrate a spectrophotometer for making quantitative measurements of absorbance, or to forget that the sensitivity of an instrument was adjusted to another level so that 'real' values can be given to the data.

F. *TAKE CARE*—At all times try to work to high standards of laboratory practice by being careful, accurate, taking account of and implementing precautions, being organised in the work, labelling material and attempting to keep errors to a minimum. In short, put into practice all the good advice given in the past about working scientifically. In addition, think and concentrate as you work and try hard to maintain consistency between trials.

G. *DO NOT INTERFERE WITH A PREPARATION WHEN IT IS SET UP FOR RECORDING RESULTS*—There is sometimes a temptation to interfere with a preparation once it is set up for recording results, however innocently; for example, there seems to be a temptation to prod animals, or to shake solutions unnecessarily while readings are being taken, or to add a bit more of a solution while a reaction is proceeding. Once such interference takes place, the conditions for recording the results have changed and the results may then be invalid.

H. *BE RECEPTIVE TO NEW PHENOMENA*—Sometimes while recording data something interesting or unexpected happens. If it does, take note of it as fully and carefully as possible and later consider whether the discovery can be usefully incorporated into the design for the remaining investigations or whether a completely new investigation should be designed to investigate it fully; for example, in a study testing the effects of amitraz (a chemical insecticide) on the feeding behaviour of neonate larvae, it was noticed that the leaf damage caused by eating the plant material was different in the presence of amitraz. Originally, plans were not made to investigate this effect but without any modification to the fundamental design useful data could be collected on this aspect of feeding.

I. *MAXIMISE THE USE OF YOUR TIME*—Generally plan the use of your time well. If you have to leave a preparation for a period of time, such as for an electrophoresis or chromatography separation, aim to use that time for something else—perhaps reading some of the literature if there is not practical work to do—but do not just sit and watch it, if unnecessary. However, if you do have to leave a preparation set up for a period like this, check that everything is running smoothly at regular intervals.

J. *PERSEVERE WHEN THINGS GO WRONG*—Invariably something will go wrong at some time or another. Persevere and try to overcome the difficulties. The more you do the more experience you will gain and the better you will get. In trying to overcome any difficulties, work systematically to trace exactly where things are going wrong and the cause of the difficulty. Invent and improvise if necessary to overcome the problem, do not just give up. If you really are at a low ebb, take a *short* break, have a coffee and come back in a better mood resolved to get something out of the practical period. It is very easy to give up but perseverance invariably pays off in the end—practical work can be quite an uphill struggle but once you are at the top of the hill travelling on from there can be easier and rewarding. There often seems to be a point in investigations when everything goes smoothly. This generally occurs when you have had some experience of the investigation and are conversant with the techniques.

4.5   A strategy for overcoming difficulties

Various aspects of what you can do to overcome difficulties are covered in chapter 5 on Results, section 5.6 (points A to E). Basically the main strategy is:

ADVICE ON FACING UP TO DIFFICULTIES

A. *IDENTIFY THE STEP WHERE THE DIFFICULTIES ARISE*—Examine the method carefully and determine at which step and at what point difficulties are arising.

B. *DETERMINE THE CAUSE OF THE DIFFICULTIES*—Examine the step or point of the method and work out the possible causes of the difficulty.

C. *MAKE ADJUSTMENTS TO OVERCOME THE DIFFICULTIES*—This will probably mean altering the design to accommodate the difficulties.

D. *PERSEVERE*—Try to be imaginative and inventive in overcoming difficulties and do not give up without an effort.

E. *WORK SYSTEMATICALLY*—At all times work systematically, going through procedures step by step, examining one feature at a time and aiming to determine the cause, or overcome the difficulty by a process of elimination—for example, check apparatus is working by going through the component parts in a systematic way, for instance check it is plugged in, set for the appropriate range, etc.

F. *DRAW ON OTHERS' EXPERIENCE*—Discuss any problems with supervisors, technicians and any others with experience. Examine the Methods Section and Discussion of the Methods in relevant papers in the literature. Others may have faced the same difficulties and found a way of overcoming them. There may even be a set of recognised procedures in overcoming them.

Very often in the initial stages of a project difficulties will arise because of your individual technique and lack of experience. For instance, in microbiological investigations the problems are often caused by poor aseptic procedures, or by a lack of familiarity with the various microorganisms so there is difficulty in identifying them; in ecological investigations difficulties are often caused by poor or inadequate sampling; in biochemistry slow, sloppy working without the necessary precautions to avoid contamination leads to problems; in animal physiology an unfamiliarity with the procedures for handling tissues to maintain function leads to unresponsive tissues; in investigations with plant material many difficulties arise from not having the correct conditions for growth, etc. As your technique improves with practice, you will probably find a number of problems are overcome. So doing the work and getting in the practice is an important aspect in overcoming difficulties.

5

Results

This chapter will help you to:

Identify what a result is

Determine what kind of results you can expect from your project

Decide when you have enough results

Analyse and interpret results

Overcome the problem of not getting any results

Use your results for choosing subsequent investigations

5.1   Introduction

Results are a desirable and expected outcome of the project work. It is therefore important that you make every effort to obtain them. Furthermore you should concentrate your efforts to produce *good* results. This chapter is concerned with looking at different aspects of results, from recognising what they are to increasing your chances of obtaining and interpreting them.

5.2   What is a result?

If you have carried out a practical investigation and been able to make qualitative observations or measurements of a variable (such as levels of certain metabolites, numbers of organisms dead, etc.) then you have results. They may not be what you wanted, they might not be good, there may not be many, nevertheless they *are* results. It is only when you have not been able to make such qualitative observations or measurements of a variable that you can say you have *no* results. Even then, you can make the statement that using that particular technique no results were obtained (this in itself might be a very useful discovery).

The results obtained can be described in different ways, both subjective and objective, as outlined below:

TYPES OF RESULTS

A. *POSITIVE OR NEGATIVE*—If a result is described as positive, it generally means that a response *was* observed. The organism *did* respond to the imposed conditions, or differences *were* found between the groups (for example, the temperature increased the heart beat rate of *Daphnia*). Negative results, on the other hand, tend to mean the opposite—that is, that a response *was not* observed. The organism *did not* respond to the imposed conditions, or differences *were not* found between the groups (for example, the herbicide did not inhibit growth).

B. *EXPECTED OR UNEXPECTED*—Quite often investigations are designed to test a specific hypothesis which has been formulated on the basis of results of previous investigations. This is basically an explanation of results and takes the form of a prediction; for instance, it might be proposed that the growth observed in coleoptiles in the presence of an auxin (growth regulator) was due to cell extension rather than cell division. The hypothesis would be that if auxin was applied to the coleoptiles the cells should be longer than the controls. The expected results would therefore be that the cells would indeed be longer than the controls when auxin was applied. Often researchers do have expected results even though a hypothesis is not formally proposed. Equally, some investigations are more 'open ended' where the outcome cannot be predicted as it is not clear what could be expected.

C. *CLEAR-CUT AND NOT CLEAR-CUT*—Clear-cut generally means that it is possible to state the result with a high degree of certainty. In a statistical analysis significance levels would be high, in other instances the results would be repeatable and consistent between trials. There are a number of reasons why results might not be clear-cut, which might be due to the method or the variability of the biological material.

D. *GOOD OR POOR*—This is basically a value (subjective) judgement on how useful or meaningful the results are. If they are meaningful then generally they can be analysed and some interpretation put on them. They have also been verified (by being shown to be reproducible). If they are useful, they generally provide some justified clarification of the original problem and/or suggest further hypotheses which can be tested by investigation. If results are poor they do not have these characteristics. On the whole, good results are produced from good investigations in which the procedures are sound, controlled and the investigations well-designed. The quality of the results therefore depends on the quality of the methods. Quite often researchers additionally see 'good results' as ones which are clear-cut. However, this is not necessarily expected in your project. Often students wrongly interpret 'good results' as ones which are positive, expected (thereby confirming an idea) and unambiguous. Rarely do they interpret good results as ones which have been verified to a reasonable extent by further or other investigations using well-designed experiments.

When you have the raw data which constitute the results, you can carry out the appropriate analyses, and from these it will be possible to make conclusions. These are brief summaries of what the results collectively show in direct relation to the aim of the investigation.

5.3   What kind of results can you expect from your project?

In some projects it might be reasonable to expect a large number of verified results, whereas in others it would be very difficult to achieve this (even though they were carried out under the same restrictions). The quality and quantity of results which can be expected from a project depend very much on the methods employed. If there are many problems inherent in the methods, or many possibilities that something could go wrong, it would be reasonable to expect relatively few results. If, however, the methods are well established, trouble free and familiar to you, many more results, and better ones in the sense that they have been verified, would be expected. You can therefore get some idea of what to expect from your project by examining the methods, using the following list to help you.

WAYS IN WHICH THE METHODS CAN AFFECT THE RESULTS

A. *THE NUMBER OF STEPS INVOLVED IN OBTAINING THE RESULTS*—There may be a number of steps which have to be carried out before results are obtained (see chapter 3, section 3.2). Each of these steps will take a certain amount of time and may involve simple or complex procedures, apparatus or instruments. The longer the time involved at each step and the more complex the procedures, the more difficult it will be to obtain results since the chances of something going wrong are increased.

Example

Consider the steps involved in examining respiration in isolated mitochondria from a marine alga: the algae have to be collected and then maintained in circulating sea water, fractionation procedures have to be carried out to isolate the mitochondria in a functional state, a respirometer (for example, a Gilson) needs to be set up under various controlled experimental conditions, and then the recordings of respiration rate have to be obtained. There are many steps and procedures which have to be completed before the results are finally collected.

B. *THE NUMBER OF PROBLEMS ASSOCIATED WITH EACH STEP*—If there are problems or difficulties associated with any of the steps, the chances of obtaining results are reduced. The more steps and the more problems, the greater the difficulty.

C. *DEPENDENCY ON SPECIALISED APPARATUS, TECHNIQUES OR INSTRUMENTS*—If obtaining results depends on a specialised instrument or apparatus, there could be problems. There may be only one piece of apparatus or one instrument within the department. There could be a high demand for it, thus limiting your access. If anything goes wrong with it you will not be able to collect results until it is fixed and if there is little expertise in using it within your department, it may take quite a time to familiarise yourself with it. If all goes well, specialised instruments and apparatus could well increase your chances of obtaining results by saving a lot of time.

The results which can be expected are therefore markedly affected by the complexity of the methods. It might be possible to enhance the quality or increase the quantity of results by adjusting the methods to simplify the procedures or reduce the number of steps. However, there is a limit to how far you can go for any particular investigation. The nature of the project will very much determine what the methods will be and therefore the type of results which can be obtained. Techniques do generally improve with practice, to speed up the process of getting results. Do therefore get as much practice as possible.

There are, of course, other factors which affect the quality and quantity of results obtained. A number of these relate to the personal characteristics of the students and include such things as: commitment; perseverance; general and specific abilities; and intellect. These characteristics determine in particular how good the design of an investigation is and how well it is carried out. (You have a limited influence on these characteristics.) Finally there is inevitably an element of luck involved in all practical investigations.

5.4   When are there enough results?

Sometimes it can be difficult to assess when you have enough results, even though you are aware of what can be reasonably expected from your project. What you should aim for is to complete satisfactorily each investigation in your project. In other words you will have enough results when each investigation has been carried out according to the design so that you can draw valid, justified conclusions from the data. (The number of investigations which should be completed can be determined from section 5.3.) The conclusions may not be what you had hoped for; for example, a significant difference might not have been found when you would have liked one, but if the investigation was carried out properly, with adequate controls and precautions and with the right amount of sampling and replication, then the investigation is complete. When one investigation has been completed others can be designed, planned and carried out. As the project proceeds you might find yourself becoming increasingly aware that you will be unable to draw valid conclusions from your work. This might occur for a variety of reasons. For example, you might not have enough data, you might have too small a sample or it may be biased; it may become evident that a crucial change in the design is necessary which had not been foreseen; or it might just be that you have been unable to complete the investigation according to the design because of a shortage of time. The fact that you will be unable to draw valid conclusions might occur even though you took great care with your original design and were conscientious in the practical work. If, therefore, you find that this is the case for your project, you will have to make the best of what you obtained. There are a number of things you can do. First identify what could be realistically expected as results from your project and see how yours compare, then see your supervisor to discuss what you have got and seek advice on the best way to make the most of what you have found. Finally, consider the possibility of doing some additional investigation with a different outlook, specifically designed to produce results quickly. The following section considers this point.

5.5 Increasing your chances of obtaining results

The advice and suggestions given in the previous chapters concentrate on showing you ways of producing worthwhile, meaningful results in your project. Sometimes, however, it becomes necessary or 'all important' to obtain results as quickly as possible, or alternatively to be certain that you will be able to obtain results on the variables you wish to study when the practical work is carried out. This might occur if you have already spent a lot of time in the project trying, but eventually failing, to get a procedure to work, or if the work you were carrying out was very original producing many unanticipated difficulties and problems which you were unable to overcome, or which were clarified too late to be investigated more thoroughly by rigorous experiment. Though such work can justifiably and usefully be presented it may well be 'bitty', and 'feel unsatisfactory'. You might prefer, therefore, to present more concrete results. However, if a lot of time for the project has passed, you will have to try to design investigations which are likely to produce results quickly. In such instances the following list identifies some of the factors you can take into account to increase your chances of obtaining results. These points develop from issues presented in section 5.3.

HOW TO GET RESULTS QUICKLY

A. *WORK ON AN ORGANISM OR USE APPARATUS WHICH IS READILY AVAILABLE, EASILY MAINTAINED AND DOES NOT INCUR HIGH COSTS*

B. *CHOOSE SIMPLE, EASY TECHNIQUES*—which are known to work reliably with the organism or type of organism you are working with and which involve little in the way of preparation.

C. *FORMULATE THE AIM PRECISELY*—(as advised in section 2.3) but keep it simple in the sense that the investigation can be carried out easily with few steps involved to collect results.

D. *INVESTIGATE MORE THAN ONE VARIABLE PER INVESTIGATION*—Try to decide on an aim where more than one variable can be observed at the same time.

E. *START THE PRACTICAL WORK AS QUICKLY AS POSSIBLE*

F. *PLAN AND ORGANISE THE PRACTICAL WORK TO MAXIMISE THE USE OF YOUR TIME*—Always check whether parallel investigations could be carried out at the same time, for example.

G. *BECOME INDEPENDENT OF ANYONE ELSE*—as quickly as possible.

H. *INVESTIGATE PROBLEMS STRAIGHT AWAY*—when anything goes wrong, analyse the causes quickly and seek help.

I. *ANALYSE AND INTERPRET RESULTS AS THEY ARE OBTAINED*

5.6   What you can do if you cannot get any results

The following suggestions should help you to identify the reasons for not getting any results and offer some suggestions of what you can do about it.

TACKLING THE PROBLEM OF 'NO RESULTS'

A. *CHECK THAT YOU REALLY HAVE NOT GOT RESULTS*—Be clear what a result is (see section 5.2). There is a tendency to think that unexpected or negative results are the same as *no* results and *THIS IS WRONG!*

B. *ESTABLISH WHETHER THE REASON YOU HAVE NO RESULTS IS DUE TO A SHORTAGE OF TIME*—You may be able to get to the stage where you were just about to collect the data but had to pack up. You will have to plan your work more carefully (see chapter 4) so that you know exactly what you will be doing before you enter the lab and how you are going to organise and carry out your work. As far as possible prepare ahead of time. You might still find it necessary to have a short extension of the practical period and should seek permission from your supervisor.

C. *FIND OUT WHERE IN THE METHOD THINGS ARE GOING WRONG*—Examine in a systematic way each of the steps in the method (see chapter 2), from obtaining the organism to collecting data, and try to pinpoint where exactly things are going wrong. For example, suppose you have got to the step where you can make measurements but that something then went wrong; by testing the apparatus you can be assured that it was working and in the appropriate range. If this was so, it would then be implied that the organism might not be functioning, and there may be ways of checking whether this is the case. You would then need to try to identify at what point the function of the organism was adversely affected.

D. *DETERMINE THE CAUSE OF THE DIFFICULTIES*—Make sure, before you look any further, that the lack of results is not caused by poor or sloppy technique, such as rough handling of the material, making up solutions incorrectly, inaccurately or carelessly, not putting into practice the necessary precautions, not concentrating etc. If this is the case make every attempt to improve by taking more care, by practising, and double-checking on the procedures with someone who understands them. Having established that there is a fault in the methods rather than in your own approach, examine carefully the step(s) you have identified as being the source of the difficulty and try to find out the exact cause so that you can make attempts to overcome it. The literature might help since you may discover that there is a small but significant change in the procedure, like for example the concentrations of certain ions in a physiological solution, or the pH, or the buffer in the extraction medium, or the strain of organism, or the timing of the investigation (daily or yearly). In addition, talk to your supervisor and others with experience in the procedures, as they may have had to face similar problems and overcome them. Some common causes for a procedure not working are: instruments not working, not working properly or in the appropriate range or sensitivity—they may not have been set up correctly or standardised in the right way; the conditions may not be close enough to the normal environment of the organism in order to maintain it in a functional state, for example there may be small changes in ion content, pH, temperature, osmotic pressure, presence or absence of metabolites or nutrients; parts of the organism may not be able to survive for a long period of time after being extracted or isolated from the organism; there may be contamination from poorly washed glassware or lack of aseptic procedures; the timing of the investigation may be inappropriate for the response being studied, for instance, the organism might normally show its response at night when you are studying it in the day; the general conditions of the lab may not be an appropriate environment in which to do the experiments—this is most evident in animal behaviour experiments; and overhandling of the material—which is a common mistake in many animal physiology experiments.

E. *OVERCOME THE PROBLEMS*—Having identified the cause of the problem you need to take steps to overcome it. Aim to eliminate it by adjusting, adapting or compromising as necessary. If you are lucky you will be able to eliminate it, by, for example, adjusting the pH of a medium to a new level, changing the buffer, avoiding overhandling etc. If, however, you have difficulties in eliminating the problem you will have to work around it in some way, particularly if you have a limited amount of time. For example, you might ideally want to work on healthy, three-week-old plants, but find that in spite of many precautions the plants are always diseased from $2\frac{1}{2}$ weeks. In such an instance, compromise is likely to be necessary by: working on younger plants; or accepting that they will be diseased and working on those with the lowest level of infection; or using a less susceptible strain. The compromise would depend on your original aim, so you would have to assess which of the factors—age, strain or healthiness was of the least importance. Again, capitalising on the experience of others can be a great help in overcoming problems, so refer to the literature or talk to your supervisor, other lecturers or technicians.

If you go through all of these points and still cannot get results in spite of adapting your techniques, seek help from your supervisor quickly, especially if you have already spent a lot of time on the work. It may be necessary to change the project's direction in order to obtain some results (see section 5.5). It is always difficult to achieve the right balance between perseverance and sensibly adjusting to the conditions of the project with the practical work.

5.7   The analysis and interpretation of results

Once results have been obtained they need to be *analysed*, in which various techniques are employed to identify as clearly as possible what exactly the results show, and *interpreted*, in which the results are explained as fully as possible, in terms of what might have occurred in the organism to produce the results shown by the analysis.

It is always useful to analyse and interpret results as they are obtained, since subsequent investigations can be designed to take account of them. It is easier to carry out appropriate analysis and interpretation if the reasons for the investigation have been clearly identified. It is worthwhile, therefore, making sure that you are clearly aware of the specific reasons for your investigation before proceeding (see section 2.5 for guidance).

5.7.1   Analysis of results

There are a number of methods available for the analysis of the results and it is likely that you have come across these in some parts of your present course. You may also find it useful to make general reference to the various statistics books listed at the end of chapter 6. Summarised below are the usual methods employed for the analysis of the results. It might be necessary to use more than one for your analyses, for example, a graph with some corresponding statistical analysis.

METHODS FOR THE ANALYSIS OF RESULTS

A. *GRAPHS*—These are useful for showing relationships between variables. Commonly the *y* axis shows the values of the variable being measured and the *x* axis the level (value) of the treatment applied within a factor, fixed by the experimenter. In this instance the latter is called the independent variable and the former the dependent variable; for example, *x* could be the temperature of water fixed by the experimenter and the *y* axis could show the enzyme activity which was measured for each of the fixed temperatures. By examining the graphs, trends and relationships (or the lack of them) can be identified—for example, a linear relationship between the variables, or a peak response identified for a particular value of the fixed variable. Ideally, you should fit the best (most appropriate, one closest to the points) line through the series of points. This requires specialist mathematical procedures, which though available through, for example, computer software packages, may not be readily available or accessible to you. Generally the simplest relationship, and the one which is easy to apply, is linear regression which fits the best straight line through the points. This can be done 'by hand' quite easily (that is, without using a computer for the analysis). It is possible to transform certain general relationships between variables into straight lines by modifying the units; for instance, an exponential curve can be transformed into a straight line by using the log of the appropriate variable, or the Lineweaver–Burk plot in enzyme kinetics for determining K_m is a transformation procedure giving a straight line. As a general guide, transform data to produce a straight line if it is possible and justified and then use linear regression to draw the best straight line through the points. If you are unable to transform the data to a straight line plot, then you will probably have to compromise and draw the curve 'free hand'. If you do this take careful note of the following. Avoid extrapolating beyond the points on the graph and if you have a number of replicate readings for each of the values of the independent variable, give the mean value and the standard error bars (see section 6.2). Take great care with interpolating (joining the points together)—each time you join points together you are making assumptions about the value of the responses for the values of x between them, therefore the more points you have (a greater range of treatment values or values of the independent variable) the more certain you will be about the curve. If there seems to be an overall trend which fitted in with the theory (such as a sigmoid dose response or growth curve) then draw the curve. If, however, there is no such trend, join the points together with straight lines between them. When drawing a curve 'free hand' take great care with peaks and troughs (maxima and minima), there is a temptation to extend the curve beyond the points to 'round it off nicely' and this rarely, if ever, is justified. Overall, it is difficult to make decisions about the type of line to draw through the points as you are beginning to place some interpretation on the results by so doing. If in doubt always seek advice.

B. *CHARTS AND HISTOGRAMS*—The commonly used charts are pie and bar charts. Each of these is useful for showing, in a diagrammatic way, relative proportions or differences. If mean values are given on bar charts, include the standard error bars or confidence limits as well (see section 6.2). A histogram is an extension of the bar chart and is useful for showing the relative occurrence of a particular value of the variable when it is arranged in order of magnitude.

C. *SCATTER DIAGRAMS*—A scatter diagram shows the relationship between two *independent* variables (neither fixed by the experimenter). If the relationship appears linear the correlation coefficient is also stated, which is a measure of the degree of association between the two variables. The points are not joined together.

D. *STATISTICS*—Statistics is a technique which is particularly useful for analysing replicated results or the results from a number of individuals in a sample. Chapter 6 deals with the procedures concerned with how to apply the statistics, in more detail (it is almost inevitable that you will need to use some of the procedures in your analyses of the results). Basically statistics allows you to:
  (i) Give precise general summaries of the variable measured for a sample—for example, mean and standard error.

(ii) Make comparisons between groups of individuals (samples), in the form of significance tests, and make conclusions. The choice of test depends on the type of sample, the measurement scale used for the variable and the number of factors fixed in the investigation.

(iii) Define the relationship between two or more variables by regression or correlation.

E. *WRITTEN SUMMARIES*—These may be necessary when you wish to draw together results in which precise measurements of a variable were not or could not be obtained, such as the general behavioural responses of an animal to a novel food stimulus, the general condition of some plants or animals at the time of the investigation, or the visual appearance and texture of a colony of microorganisms. The summaries should be as precise and quantitative as possible, so for example try to give some measure of how commonly a characteristic occurred: for example, in *all* cases the organism . . ., or in 10 per cent of the animals displacement activity in the form of cleaning the whiskers was observed . . . etc. Your main aim here is to try to identify trends and anomalies by careful consideration of the results.

Having analysed the results it is useful to write a series of statements from the analyses (conclusions) which indicate, in very clear terms, what has been shown. For instance: there was a significant difference in the mean weight of the limpets taken from an exposed and a sheltered shore; there was a peak absorbance of the pigment extracted in water from the alga at 620 nm. If you had been applying some experimental conditions, make sure your analyses indicate clearly the effect of those conditions. For instance: an increase in temperature from 10°C to 20°C did not alter the enzyme activity; extract of the thoracic glands produced moulting in all cases when applied after the animal had had a meal.

Overall be honest in the statements you make for yourself about the results—identify what really happened not what you would have liked to have happened. Students very often do not recognise or appreciate the value and importance of being honest, and they spoil their work by 'fixing' the results to what they think should have happened. It is often the unexpected results which lead to new insights and ideas for further investigation.

5.7.2   Interpretation of results

In interpreting results you will be attempting to explain exactly what happened in the organism to cause it to respond in the way it did— that is, you might be trying to explain *exactly* how the drug could have acted to decrease the foraging activity of mice; what might have happened to prevent the pathogen from increasing the disease incidence compared with the controls; why exactly the growth characteristics of the grass *Molinia* should be different when regularly grazed. In the formal sense you will be attempting to propose hypotheses to explain your results. To interpret results means that the methods must be examined first, in order to establish that the procedures employed did not themselves explain the results (for example, the low activity of an enzyme may have been caused by inactivation due to a low pH in the final extraction, rather than because it had naturally low levels in the tissue being studied). The biology of the organism must then be examined in detail, in relation to the conditions of your investigation, to try to find an interpretation. It can be difficult to interpret results. Probably the best way to gain an understanding of what approaches are taken is to read the Discussion sections of the research papers published in the journals. These put into practice many of the points referred to below, which should provide you with some guidance on how to approach the interpretation of results.

A GUIDE TO INTERPRETING RESULTS

A. *MAKE SURE THAT YOU HAVE ANALYSED YOUR RESULTS AND THE ANALYSIS IS CORRECT*—Having analysed the results you should have clear statements of what you have found and therefore what you need to explain. If you have not analysed the results it is difficult to know therefore what needs to be interpreted, and if the analysis is incorrect, inappropriate or wrong then the interpretation will be a false explanation of what was really happening in the organism. Common mistakes made in analyses include misreading graphs (because a result is wanted or expected), using the wrong statistical test, and ignoring the outcome of a statistical significance test (for instance, suppose the test showed that there was no difference between two groups, the tendency is to ignore it when a significant difference was wanted and to say that there is really a difference between the groups).

B. *CONFIRM THAT THE RESULTS ARE NOT DUE TO A 'QUIRK' IN THE METHODS*—Sometimes the methods can affect the results in a direct way and in ways not always evident at the outset. Some common ones include biased sampling, poor matching of samples, too small samples for the method of analyses or for the nature of the investigation, extraneous factors operating during the investigation, inadequate controls, bias in the collection of the results (such as ignoring results you do not like the look of), inadequate precautions, and inappropriate or not correctly applied measurement techniques.

Example

A student was comparing Na^+ excretion in urine of smokers and non-smokers. An aliquot of urine was collected in the same way from matched samples and Na^+ levels measured with a flame photometer. Na^+ levels were significantly different between the two groups and less in the smokers. It appeared to the student that the smokers were excreting less Na^+ per day than the non-smokers. However, the flame photometer measures *concentrations* of Na^+ and the student took no account of the volumes of urine produced. It was therefore difficult to deduce whether Na^+ excretion was indeed less in the smokers. Subsequent experiments properly controlled showed that the daily amount of Na^+ excreted was the same but the volume of urine produced differed. The first results obtained were therefore due to a 'quirk' in the methods. The interpretation was affected because in the first experiments explanations were sought to explain why the Na^+ levels differed in the two groups whereas in the second, better-controlled experiment, the physiology was studied to try to explain why the *volume* had changed (and not the daily Na^+ excretion rate).

C. *TAKE ACCOUNT OF ANY LIMITATIONS IMPOSED BY THE METHOD IN THE BIOLOGICAL INTERPRETATION*—You may find that the method you used was limited in some way. For example, you could not sample from the population planned and had to be more selective in sampling, or your final sample size was very small, or the particular instrument used had a limited range of response or was inaccurate beyond certain maximum or minimum levels. If such factors are evident in your investigations when collecting your data, you must take account of them in interpreting the results and not ignore them.

D. *FAMILIARISE YOURSELF WITH THE BACKGROUND TO YOUR INVESTIGATION*—Since the work you are doing is probably an extension of previous work, knowledge of the background is likely to give you ideas for the explanation of your results. For instance, your investigation may be to test a particular hypothesis proposed from research papers in the literature. The hypothesis is a prediction of what might have happened if one chosen explanation is correct and that explanation will be given in the paper proposing the hypothesis. You should then be able to use the explanation in interpreting your results (whether the hypothesis is supported or not). An alternative might be that your investigation may involve the attempted confirmation of previous studies. If your results are similar to those reported, it is likely that the explanation for your results is the same as for those reported. If your results are different, the explanation will have to be altered and might even be the opposite to that proposed.

E. *FIND OUT AS MUCH AS YOU CAN ABOUT THE VARIABLE YOU ARE INVESTIGATING AND WITHIN THE CONTEXT OF YOUR INVESTIGATION*—By doing this you will increase your chances of coming up with sensible explanations. You need to consider such aspects as: what is the variable, how is it explained, what are its general characteristics, what are its particular characteristics in the system you are working with, what kinds of factors affect it and in what way, what are its specific characteristics in similar and different organisms/systems environments etc. and how are these explained?

Example

Suppose the investigation was to test the effect of nitrate concentration on nitrate uptake by cereal plants. The kind of questions you could ask are what are the possible uptake mechanisms, what are their characteristics, how are they distinguished, how are they affected by concentration, what is known about nitrate uptake by plants in general, in other organisms, or cells, is there anything particular about nitrate and cereals or cereals?

F. *APPLY THE KNOWLEDGE GAINED IN PROPOSING EXPLANATIONS FOR YOUR RESULTS*—With the results summaries in front of you, the information about the variable and the background of the project, begin to formulate and examine various explanations of the results. Do not try to explain every small change observed but look initially at overall trends and effects and then look in more detail at 'odd' results in the light of the trends and overall effects identified. For example, many students attempt to explain every small rise and fall in a graph and not notice, for example, the overall linear relationship between certain variables, or they concentrate on differences in individuals which make up a sample and do not take clear enough note of the fact that significant differences occurred between the samples in spite of the variation within the sample. Having obtained a clear idea of what results have been obtained you now need to propose the explanation clearly, examine it in depth and avoid being superficial in explaining the results. Firstly, clarify exactly why and how the explanation suffices for your results (there is a tendency among students not to do this and instead merely state the explanation without examining it in depth). What this means is that you will have to look very closely at your organism and try to visualise each of the steps which must have occurred to give the results obtained. You can check whether such steps are likely or feasible for your organism. What you will be doing is thinking through the argument for the explanation as thoroughly as possible to conclusion—for example, determining each possible step that led to an increase in nitrate uptake from solution to cell. While doing this, collect together information which both supports and does not support your explanations, always being clear of the reasons why it does either, and from this make some assessment of how likely it is that the explanation is correct. Be honest! Do not make the mistake of claiming too much from relatively few results, and try to recognise any limitations of your work; for example, do not claim to have solved the problem of elucidating the mechanisms of nitrate uptake of plants from your restricted studies. In order to test how good your explanation is try to make predictions from it. If the explanation is good it should be possible to make predictions which can be verified, either by further study or from data already available in the scientific literature. For example, if nitrate uptake was due to diffusion it should be possible to show that the process was temperature, pressure and concentration dependent.

G. *USE EACH OPPORTUNITY AVAILABLE TO DISCUSS INTERPRETATIONS*—Discussions with your supervisor and seminars arranged to discuss the project work provide opportunities to discuss the interpretations of your work and can therefore be a source of new ideas and insights to the work. It is important, therefore, that you do make sure of taking advantage of such opportunities to discuss your results.

5.8   Using the results for subsequent investigations

When you have the results of your investigation analysed and interpreted, you will be in a position to consider subsequent investigations. You should have some idea of how much time you have left and whether you need to do further work. When you were preparing your project initially and chose an investigation to start with, you should also have had some appreciation of how it might develop into parallel or serial investigations (see chapter 2). If you had planned parallel investigations, see whether outstanding ones are still appropriate and useful in the light of your results and the overall aim(s). Consider, too, whether it might make more sense to continue with a serial investigation. A useful way to proceed with serial investigations is to make a list of questions and ideas that you have had after interpreting the results. In attempting to explain your results you are likely to have proposed that the results obtained may have been due to 'X' or 'Y'. Subsequent investigations could usefully test whether the explanation is likely. In other words you will be aiming to obtain more evidence to substantiate or not (as the case may be) your explanation.

On the whole the investigations which follow the first may fall into one or two types. They might be related to the methods. You may have realised or suspected that some aspect of the methods would be influencing or biasing the results. It would make sense, therefore, to investigate this more fully.

Example

Suppose that you suspected that the growth medium was favouring the growth of certain microorganisms, giving biased results, or that the method of fixation caused a marked change in the tissue. In such instances, modification of the methods could clarify your suspicions.

Alternatively you may feel that the methods are sound and that there is no real justification for investigating them further. The subsequent investigation could therefore be designed to amass further evidence for an explanation or to study alternative and additional hypotheses.

Example

Suppose you had evidence to suggest that acetylcholine was a transmitter at a particular synapse. Subsequent investigations could study more fully the effects of agonists and antagonists etc., to build up the evidence for this view point.

Example

Suppose the results of your experiments together with evidence from the literature led you to hypothesise that the activity observed was governed by a biological clock. Subsequent investigations could begin to examine this hypothesis in detail.

With various ideas on what to proceed with, go back to chapter 2 to choose an investigation from the list, and define the aim precisely (sections 2.4, 2.5 and 2.6).

6

Using and Applying Statistics

This chapter will help you to:

Identify what statistical techniques are commonly used for biological data analysis

Recognise the different applications of various techniques

Apply appropriate statistical analysis to your data

Appreciate the common principles on which significance tests are based

Avoid obvious pitfalls in applying and using the various statistical techniques

6.1   A basic approach to using and applying statistics

Using statistics is like using any other technique for helping you in an investigation. You must know: what the technique is useful for; what you specifically want to use it for and the rules for using it, so that you can apply the techniques correctly and appropriately. Statistics is extremely useful for helping you to analyse data obtained from investigations in which there has been some kind of replication (such as samples) and for helping you in the construction of graphs. It can often be helpful in planning the design of investigations. Students commonly have an exaggerated fear of statistics, probably brought on by a fear of 'maths' or unhappy experiences in the 'maths and stats' lectures. The most important thing to grasp initially is when and how to apply the variety of statistical techniques. If this is grasped, any fear present begins to disappear as confidence takes over.

On the whole, the statistical techniques which are readily available and generally needed for the analysis of data from biology projects fall into the categories outlined below. An appreciation of this can be a great help to you in the application of statistics.

CATEGORIES OF STATISTICAL TECHNIQUES

A. *SUMMARIES OF DATA (section 6.2)*—These provide a concise and precise way of summarising quantitative data obtained from a sample or by replication. For example, it is common to give the mean as an average value and the standard deviation as an indication of the variability of the results.

B. *STATISTICAL SIGNIFICANCE TESTS FOR COMPARING ONE SET OF DATA WITH ANOTHER (sections 6.3, 6.4 and 6.5)*—These tests provide a means of making comparisons between one group of data and one or other groups in order to determine if there are significant differences between them or not. If significant differences are found, the level of significance can be stated. For example, the 't' test is a significance test which compares two means to see if they are different.

C. *DETERMINATION OF THE RELATIONSHIPS BETWEEN TWO VARIABLES (section 6.7)*—These are techniques used when any individual in a group is defined by two measured variables rather than one and the relationship between the two is defined. One variable may be fixed, as with experimental treatments such as a range of pH or temperature.

(i) CORRELATION (section 6.7)—This is a technique which expresses the degree to which two variables, measured independently on each individual in a group or population, vary together in a linear way. In other words it is a technique for expressing the degree of correlation between two variables by working out the correlation coefficient. For example, the length and width of an insect wing was studied and the correlation coefficient was 0.9, indicating high correlation. If it had

been zero there would be no correlation and thus the two variables would be independent of one another.

(ii) FITTING THE 'BEST' STRAIGHT LINE OR CURVE ON A GRAPH (section 6.7)—This is a technique called regression analysis whereby the 'best' straight line or curve is fitted to a series of points on a graph. It therefore defines the relationship between the points on a graph in a very precise way.

Within each category there are then different techniques available which have to be matched to the measurement scale of the data to be analysed (see chapter 3, section 3.4, for explanation of scales), and the specific reasons for the analysis. In choosing a technique, therefore, an essential first step is to be absolutely clear what you want to use the statistical techniques for. (This should be evident from the reasons for an investigation identified in formulating the specific aims—see chapter 2, section 2.5.) For your own investigations, work through the following flow diagram to identify what statistics you wish to apply. You may find that you need to use more than one category to satisfy

FLOW DIAGRAM FOR CHOOSING A STATISTICAL TECHNIQUE

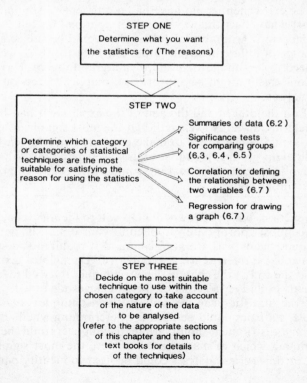

STEP ONE
Determine what you want
the statistics for (The reasons)

STEP TWO

Determine which category
or categories of statistical
techniques are the most
suitable for satisfying the
reason for using the statistics

Summaries of data (6.2)

Significance tests
for comparing groups
(6.3 , 6.4 , 6.5)

Correlation for defining
the relationship between
two variables (6.7)

Regression for drawing
a graph (6.7)

STEP THREE
Decide on the most suitable
technique to use within the
chosen category to take account
of the nature of the data
to be analysed
(refer to the appropriate sections
of this chapter and then to
text books for details
of the techniques)

the aims. For example, you may wish first to determine the value of a variable and then compare it with another group or population. The flow diagram summarises what you need to do to choose an appropriate statistical technique.

Examples

1. Suppose the aim of the investigation was to compare the cell length of coleoptile epidermal cells grown in the light and dark. The data would first have to be summarised, so it was clear what length the cells were (on average), and then comparisons would need to be made between the two groups of coleoptiles. Statistical techniques for summarising data and making comparisons would therefore be used.

2. It might also have been an aim of the experimenter to see if any relationship existed between total coleoptile extension and cell size. In this instance correlation could be used for each of the experimental conditions (light and dark) to see if there was any relationship between the extension of the coleoptile and the cell length.

3. The experiment could have had a different reason, aimed to test what effect the wavelength of light had on cell extension. A number of coleoptiles would be subjected to a range of different wavelengths. The data for these would need to be summarised and then a graph drawn. Having done this it might be appropriate to apply regression analysis.

4. If on the other hand the aim of the experiment had been to determine what wavelength had the most significant effect, it would be necessary to make comparisons between the different groups, grown at each wavelength, and so significance tests for making comparisons between groups would be chosen.

The sections which follow will help you to identify what technique could be most appropriately applied to your data. The details of the techniques themselves are not given and if required statistics books will have to be referred to. A number of very helpful textbooks are listed at the end of the chapter. The techniques referred to here are by no means exclusive in terms of the categories identified, the uses to which the statistics can be put or the techniques available. For example, statistics could be used for determining whether data were randomly distributed, multivariate techniques could be used for determining which of many variables were the most significant in a study, or time series analyses could be used to identify patterns over time. Such techniques are available in more advanced texts, though some are covered in the books listed at the end of the chapter.

There are also a number of computer software packages dealing with statistics commercially available for both micro and mainframe computers; for instance, Minitab and SPSSX (Statistical Package for the Social Sciences) are two comprehensive statistics packages available for mainframe computers. Some lecturers in your department may even have written their own programs that you can use. When you have your data it is generally best to open a file in the computer to store them. You can then draw on the data as you need to carry out the necessary statistics.

6.2 Statistical techniques for summarising data

In summarising data it is advisable to give the following.

USUAL COMPONENTS IN A SUMMARY OF DATA

A. *AN 'AVERAGE' MEASUREMENT* (in statistical terms a measure of central tendency) — Usually the mean, median or mode.

B. *A MEASURE OF HOW MUCH THE DATA VARY* (in statistical terms a measure of dispersion) — There are a number but the commonly used ones are the standard deviation, the standard error, the range and the confidence limits. The variance and the coefficient of variation are others.

The technique chosen is matched initially to the measurement scale of the data. You therefore should recognise what measurement scale has been used for your data (see chapter 3, section 3.4) and refer to the appropriate point below.

6.2.1 Ratio and interval scale data

A useful way to summarise this kind of data is to give the mean ± the standard error and the sample size; for example, the mean muscle tension was 0.5 ± 0.1 g (± standard error), $n = 7$. It is always important to identify what kind of summary term is being used in such a final statement. The standard error gives enough information for other measures of dispersion to be calculated as long as the sample size is also given; and a summary in this general form is sensible for making comparisons with other data, particularly when parametric statistical tests are to be used (see section 6.4). When giving mean values on a graph, it is also necessary to give a measure of dispersion, and normally standard error bars or confidence limits are given as follows.

In either case it should be made clear what exactly is being shown; that is

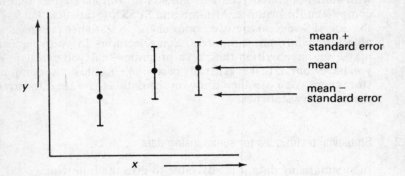

The median and mode are useful for summarising data which are asymmetrically (rather than normally) distributed. The mode is particularly useful if there is more than one peak on a frequency distribution (such as a histogram) as each mode can be identified and the distribution is identified as being multimodal.

The coefficient of variation is a particularly useful statistic for summarising the variation in a population and for making comparisons between the amount of variation in populations having different means. (It is defined as $(s \times 100)/\bar{x}$, where $s =$ the standard deviation and $\bar{x}$ is the mean.)

6.2.2   Ordinal (or rank) scale data

The median with the range and sample size is the most useful summary for this kind of data and particularly for when non-parametric statistical tests are subsequently used (see section 6.4).

6.2.3   Nominal scale data

These data are already usefully summarised in table form. (*Note:* Sokal and Rohlf (1969) give a very useful discussion of the relative merits of these summary statistics in chapter 4 of their book.)

6.3   Significance tests for making comparisons between sets of data to test for significant differences

The text which follows is concerned primarily with tests which test for significant differences between samples from defined populations. There are additional statistical significance tests for making other kinds of comparison, such as for comparing correlation or regression coefficients, or testing whether a sample belonged to a population with known population parameters. The tests which are covered here are shown in table 6.1.

6.3.1   Choosing a statistical significance test

BASIC PROCEDURE FOR CHOOSING A SIGNIFICANCE TEST

A. *DECIDE WHICH TYPE OF STATISTICS IS TO BE USED*—Parametric or non-parametric.

B. *IF NON-PARAMETRIC, MATCH THE MEASUREMENT SCALE OF THE DATA TO THE TYPE OF STATISTICS*—For example, ordinal scale data uses rank order statistics tests, nominal scale data uses others.

C. *DECIDE WHICH SPECIFIC TEST SHOULD BE USED*—Take account of what you want from the test; the number of samples to be compared; the number of experimental factors involved.

6.3.2   Deciding between non-parametric and parametric statistics

Table 6.2 summarises which type of statistics should be used for each scale of measurement. A fuller discussion is given in the text which follows in this section.

Parametric significance tests make certain assumptions about particular characteristics of the population under study called parameters (such as the mean, μ, the standard deviation, σ, the variance, σ^2, of the population—all denoted by Greek letters). They are used for interval and ratio data and make comparisons between means or variances to test for significant differences.

Non-parametric significance tests, on the other hand, do *not* make assumptions about the population parameters and are used for ordinal and nominal data and for interval and ratio data when the assumptions of the parametric tests are seriously violated (often called 'distribution free statistics' therefore). Non-parametric tests for ordinal data are called 'rank order statistics' and compare groups for overall differences (there are a number of non-parametric rank order statistics tests which are equivalent to the parametric tests—see table 6.3). Non-parametric tests for nominal data, on the other hand, compare frequencies (numbers of individuals having a particular characteristic) or proportions for significant differences.

Table 6.1 Some general information on significance tests

Significance test	Function	Conditions	Null Hypothesis (N.H.)	Standard unit	Assumptions	Useful points
t test Paired t test	Compares two sample means for significant differences	Samples are paired. Interval or ratio data		t	Random samples taken from normally distributed populations with equal variances	t tests are robust tests and can be used in most cases unless the assumptions are seriously violated
Unpaired t test		Samples are independent. Interval or ratio data	There are no significant differences between the means	t		
ANOVA One way ANOVA	Compares two or more sample means for significant differences	Samples are independent. Interval or ratio data. Each sample is subjected to a different level of treatment within one experimental factor		F	General assumptions: (i) effects are additive (ii) random samples come from normally distributed populations with equal variance	For all ANOVA tests, further analysis can be carried out to test between which specific means significant differences occur. In two way ANOVA without replication, it is generally not recommended to test for significant differences between the matched individuals as well.
Two way ANOVA without replication	Compares two or more sample means when each sample is subject	Samples are matched. Interval or ratio data		F		

PARAMETRIC STATISTICAL SIGNIFICANCE TESTS

with replication	for each of two experimental factors and tests for interaction between the factors	two experimental factors and a range of levels within each. Each sample therefore is subjected to two levels of factor A at the same time, one belonging to experimental factor A, one to experimental factor B. The treatment identifies exactly what combination of levels of the factors the individual receives	1. No significant difference between treatment means of factor A 2. No significant difference between treatment means of factor B 3. No significant interaction between A and B	F	in missing values. If significant interaction occurs, re-analyse using a series of one way ANOVA. Check on the model of the ANOVA (I, II or III) for analysing F values for each of the Null Hypotheses
t test equivalents Wilcoxon	Compares two samples for overall significant differences	Samples paired. At least ordinal data	Samples from populations having the same distributions— that is, there is no significant difference between the samples	T	
Mann–Whitney		Samples independent. At least ordinal data		U	No assumptions made about population parameters Take care using tables of U. Some books give the largest value and some the smallest

NON-PARAMETRIC TESTS

(continued on page 90)

Significance test	Function	Conditions	Null Hypothesis (N.H.) unit	Standard	Assumptions	Useful points
ANOVA equivalents Kruskal–Wallis	Compares more than two samples for overall significant differences between treatments	Samples independent each subjected to a different treatment. At least ordinal data	Samples from populations having the same distributions—that is, there is no significant difference between the samples	H	No assumptions made about population parameters	
Friedman		Samples matched subjected to different levels of treatment. At least ordinal data		χ^2		Kendals coefficient of concordance may be used to look at the significance of differences between matched individuals
Chi square Goodness of fit	Compares observed frequencies with those expected (from a theory) for one sample	Nominal data. One sample	Observed frequencies equal expected frequencies	χ^2	Categories are mutually exclusive and totally inclusive	The expected cell frequencies should not generally be less than five for any category. Yates correction should be used in contingency tables with only one degree of freedom.
Contingency table	Tests for independence or association between two simultaneous classifications	Nominal data. A table of frequencies is constructed in which individuals are classified by	There is no association between the two classifications (factors); they are independent			Ensure all individuals classified are included in the table before analysis

Table 6.2

To show which type of statistics is appropriate for the measurement scales

	Statistics	*Scale of measurement for data*
A.	Parametric	Interval or ratio
B.	Non-parametric	
(i)	Rank order	Ordinal, interval or ratio when assumptions seriously violated
(ii)	Non-rank order	Nominal

For interval or ratio data it is clear that either type of statistics can be used. Basically, you should always use parametric tests in preference to non-parametric tests since they are more powerful and efficient and make best use of all the available data. However, the parametric tests should not be used if the assumptions are seriously deviated from. It can be difficult to assess this. In the parametric significance tests which make comparisons between the means ('t' test and Analysis of Variance), common assumptions are that the samples from which the means have been calculated come from normal distributions with equal variances. It is possible to test whether variances are significantly different by using an 'F' test. However, the test will be based on limited data from the samples and it is therefore questionable how useful it is. However, Zar (1974) points out that there have been a number of studies which indicate that the 't' test (which makes comparisons between two means) is robust enough to stand considerable departure from the theoretical assumptions, especially if the following conditions are adhered to in the experimental design and analysis: sample sizes are equal or kept nearly equal; two tailed tests are used in preference to one tailed tests, the tests do not depend on very small significance levels, that is, 0.01. Similarly, the ANOVA (which compares two or more means) is a robust test and can withstand deviations from the assumptions. Keeping sample sizes equal or nearly equal will help satisfy some of the assumptions. Denenberg (1976) also shows that by keeping to samples of 20–30, the assumptions of the 't' test become relatively unimportant.

Certain common types of data do not follow a normal distribution but can be easily transformed into a form that does, and parametric tests can therefore be used.

In trying to decide what statistics to use therefore, take these points into account and: generally use parametric tests; put into practice the recommendations by Zar (1974) and Denenberg (1976) if possible; and transform data which are known *not* to follow a normal distribution.

Examples

(i) Percentage data: use the arcsin transformation (for example, percentage increase in length of root).*

(ii) When s (the standard deviation) is proportional to $\bar{x}$ (the mean), that is, a constant coefficient of variation (growth data are commonly of this form): use the logarithmic transformation.*

(iii) When s^2 (the variance) is proportional to $\bar{x}$ (the mean), which is usual for data following a Poisson distribution: use the square root transformation.*

(* Details relating to these transformations are given in Zar (1974) and other statistical books.)

6.3.3   Choosing the specific significance test for your data

Having decided what type of statistics should be applied to your data, you then have to make a choice of the specific test. Section 6.4 considers the parametric and rank order tests and section 6.5 the test to be used for nominal data. The appropriate section should now be referred to. Table 6.1 gives a summary of the commonly used tests.

6.4   Parametric tests for comparing means and their non-parametric, rank order statistics, equivalent

There are two types of significance test for comparing means: the 't' test for comparing *two* means and the Analysis of Variance (ANOVA) for making comparisons between *two or more* means. There are non-parametric equivalents of these in terms of the number of samples compared and these are shown in table 6.3. However, the non-parametric tests test for overall differences between the samples rather than for differences between the sample means specifically. 't' tests and ANOVA are further subdivided into various types which match certain experimental designs. The main ones are summarised in table 6.3 (though there are others). The table can be used for deciding which specific test is appropriate for your data. Explanations of the terms used are given in the text which follows the table and some examples of the tests are given in table 6.4.

Table 6.3 For identifying a specific statistical test for your data

Type of statistics test	Sample type	Significance tests	
		2 samples only	2 or more samples*
Parametric (compares means)	Matched samples	Paired t test ⟶	*Two way ANOVA without replication— model I, II or III (preferably II or III)
	Independent samples	Unpaired t test ⟶	*One way ANOVA—(when only one experimental factor is varied)—model I or II, usually model I *Two way ANOVA with replication (when two experimental factors are varied and two sets of means are compared— one set for each factor)— model I, II or III
Rank order non-parametric (makes comparisons for overall differences)	Matched samples	Wilcoxon matched pairs test	*Friedman's two way ANOVA test
	Independent samples	Mann–Whitney U test ⟶	*Kruskal–Wallis one way ANOVA test

Key ANOVA = Analysis of Variance.

   *The tests compare for overall differences between averages. Multiple range testing can subsequently be used for testing precisely between which averages significant differences occur, for example, the Newman–Keuls test.

Model I = fixed effects model; II = random effects model; III = mixed effects model.

Some definitions for using table 6.3

A. The type of samples to be compared (see chapter 3, section 3.5, on samples, point D) There are two types of samples, matched or independent. Matched samples may be considered as samples in which each individual measurement for one sample is related (or matched) to one individual measurement of every other sample.

   Suppose, for example, there were three samples, A, B and C, and for each sample there were four individual measurements. Then for a matched sample individual number 1 in A would be related (or matched) to individual number 1 in B and C. Individual 2 in A would be matched with individual 2 in B and C and likewise for individuals 3 and 4. Ideal matching is often obtained by using the same individual for A, B and C or litter mates, but individuals may be matched according to their genetic composition, weight, previous experience, etc. If there are only two samples, the samples are often described as being paired. Independent samples in comparison may be described as samples in which each individual measurement of a sample is unrelated or independent of any individual measurement in any other sample (compare with matched samples). If there are two samples, the samples are often referred to as unpaired.

B. The number of experimental factors or conditions altered within the investigation (see also chapter 2, particularly section 2.6.2)

In an investigation there is at least one factor which is evaluated in which comparisons are made between samples subjected to different treatments. For example, the factor may be temperature and comparisons may be made between the results at two or more different temperatures (treatments), or the factor could be a species where results were compared for two or more different species (treatments), or the factor could be a herbicide where results were compared in the presence or absence of the herbicide. The investigation thus compares experimental results with a control or between herbicides at different concentrations and a control. The two way ANOVA and the Friedman non-parametric test can be used for an investigation in which two factors are identified in one investigation. The aim would be to evaluate the effect of each of the factors on the results. In a two way analysis of variance with replication, any combined effects of the two factors (called interaction) can also be tested for. Table 6.1 shows the null hypothesis tested which reflects this, and the following diagram may help which shows the basic experimental designs.

(i) *Investigations with one factor*   (ii) *Investigations with two factors*

C. Models of the Analysis of Variance

Model I—Fixed treatment effects model. The treatments are fixed and determined by the experimenter. The treatment is therefore repeatable, such as fixing a range of concentrations, or identifying a particular population, species or strain from which to sample and to make comparisons.

Model II—Random effects (or Variance component model). The treatments are not clearly fixed and determined by the experimenter and therefore repeatable. For example, suppose different mice were analysed for protein levels of heart muscle cells. Four mice may be taken and three estimates of protein made on each. The experimenter may wish to determine if there were differences between the mice. The mice were not treated in any different way and were taken randomly from a population. The exact conditions of each of the treatments, that is, the mice, could not therefore be repeated. The exact conditions of the treatments were not therefore under the experimenter's control and occurred at random. The model is thus a model II design.

Model III—Mixed effects model. This occurs in two way analyses of variance where one factor may be a fixed treatment (model I) and one random (model II).

Discrimination between the models is particularly important in the computation for the two way analysis of variance for determining where significant differences occur.

Table 6.4

Some examples of the significance test shown in table 6.3

Parametric tests

Paired t test	To test if the daily excretion rate of Na^+ is altered in male mice after receiving an injection of aldosterone (the test is a paired test because the same animals were used in the control (before) and the experimental (after) situation)
Unpaired t test	To test if there is a significant difference in the nitrogen content of two soils taken from different locations
Analysis of Variance, one way	To test whether there is a significant difference in the yield of bacteria (measured as numbers of bacteria) grown in different media for 13 days (Liver digest, minimal (salts plus glucose), Nutrient broth, tryptone)
Analysis of Variance, two way without replication	To test if there was a significant difference in the amount eaten by rats when given a choice of three different types of food. (It is a matched test because the same animal was given the choice. A number of different animals were used, each given a choice, thereby making up the sample)
Analysis of Variance, two way with replication	To test if the vitamin C content varies with the type of fruit and the storage conditions. The two factors are types of fruit and storage conditions. A sample of fruits was analysed for each of the treatments (that is, for any particular fruit and storage condition). This is a type I design because both sets of factors are fixed by the experimenter

Non-parametric tests

Wilcoxon	To test whether the amount of displacement activity shown by male monkeys increased when the animal was placed in an impoverished

(continued)

	environment in which there was little sensory stimulation. (The same animal was used in the deprived and enriched (lots of sensory stimulation) environments. A method was devised for giving a measure of displacement activity using an ordinal scale of measurement)
Mann–Whitney	To test if the flower colour of a particular species of plant differed for plants of two populations from different locations. (A scheme was devised for giving a colour measurement which used an ordinal scale of measurement)
Kruskal–Wallis	To test whether the degree of infection was significantly different between four strains of pea plant. The degree of infection used an ordinal scale of measurement
Friedman	To test if the degree of aggression shown by pigs is significantly different after being reared in different conditions. A scheme was devised for making a measure of aggression on an ordinal scale. Pigs were matched across the different conditions by sex, size and litter

6.5   Chi square tests for testing for significant differences with nominal data (χ^2)

There are a variety of significance tests for analysing nominal data and Siegel (1956) and Sokal and Rohlf (1969) cover a number of these. However, in this chapter, only Chi square will be considered in two instances: firstly with one sample where a test of goodness of fit of the nominal scale data is tested against frequencies predicted from a theory and secondly for contingency tables where the Chi square test tests the association between two categories used to classify the data.

6.5.1   Chi square goodness of fit test

This test is used on nominal data collected from one sample. Frequencies are therefore placed in mutually exclusive classes and the test is then carried out to see if the frequencies obtained fit those expected from a particular theory. It is commonly applied in genetics experiments where it is used to test whether the ratio of the different phenotypes of offspring, resulting from a genetics cross, fit a theoretical ratio like 3:1, 9:3:3:1 or 1:1:1:1 for example. However, there are other uses whenever predictions can be made about what frequencies to expect.

Examples

You may wish to test whether some organism was randomly distributed in space or time. If it were, it would follow frequencies per unit area (for example, number of parasites per host, number of plants per quadrat) according to a poisson distribution. Another instance may be where you wish to propose in an experiment that there was an equal likelihood that an individual would be in a defined space at any one time. For example, a rat may have a choice of three foods. You could hypothesise that the rat showed no preference and therefore was equally likely to eat each of the foods. If you took a sample of rats, therefore, you would expect the frequencies to follow a 1:1:1 ratio if the rats were showing no choice or preference for the different foods.

6.5.2   Chi square contingency tables

Contingency tables can be produced when the frequencies observed can be classified by more than one set of criteria and thus placed in a table of the following type which shows two categories of classification, A and B.

category A

	A1	A2	A3	A4	A5	etc.
B1						
B2		Frequencies placed in				
B3		the appropriate cell				
B4						
etc.						

category B

For example, A could be eye colour and B hair colour. A1–A5 and B1–B5 are classes which represent different eye colours and hair colours respectively.

The Chi square test tests the degree of association between the two categories; or another way of looking at it is that it tests whether the two categories are 'independent of one another', that is, are not associated. The categories can be defined in any way; they could be samples, experiments, species, days, etc. However, as is typical with nominal data, the classes must be totally inclusive and mutually exclusive within the category. This test can be used with proportions and percentages, but the figures in the contingency table must be in the form of frequencies or numbers of individuals as before. The general format of the table is

		category A				
		A1	A2	A3	A4	etc.
category B	+					
	−					

Here the + identifies the number of individuals giving the required response and the − indicates the number not giving the response.

Examples

The + and − could represent the number of seeds germinating or not respectively, under five different concentrations of copper in the growing medium (A1–A5), or could indicate the number of individuals in a sample who contracted a disease when either they were vaccinated or not vaccinated (A1–A2). If figures were originally in terms of per cent, for example 6 per cent were infected, the number 6 would be in +, and 94 in −.

The important thing to remember in using such a table is to make sure that the frequency total equals the number of individuals investigated. There is a tendency when producing such a table to include only the + frequencies, or just the − frequencies, but this is incorrect.

This χ^2 contingency table analysis can be a very useful test in biology, for example, allowing tests of homogeneity to be made between experiments or recorders, or for comparing samples of nominal scale data. There are many variations (and manipulations) of this test, each with particular assumptions; see the texts quoted at the end of the chapter for details.

6.6 Understanding how to use a significance test

In order to understand the procedure for carrying out a statistical test, and what it means when it has been completed, it is useful to recognise that significance tests follow a common procedure.

COMMON PROCEDURE IN A SIGNIFICANCE TEST

A. *A NULL HYPOTHESIS IS SET UP*—Using a standard hypothesis for the test.

B. *THE NULL HYPOTHESIS IS TESTED*—Using a standard statistical test.

C. *CONCLUSIONS ARE MADE AS A RESULT OF TESTING THE HYPOTHESIS*

The stages are expanded on in the text which follows.

A Stage one—Set up the Null Hypothesis

The Null Hypothesis is fixed by the statistics test and is set up without reference to the observed data. It cannot be changed at will, since the rest of the procedure depends upon the fixed Null Hypothesis. For example, the Null Hypothesis for the t test states that there is no significant difference between the means. Of course, you can translate this into the biology and it is worthwhile doing so since you will probably then have a clearer idea of exactly what you are doing, by for example stating that there is no significant difference in the mean heart rate before and after exercise.

You may need to check on the format of the Null Hypothesis for your test, but to help you a list of the significance tests along with the Null Hypothesis or Null Hypotheses they test are shown in table 6.1. An alternative hypothesis is also set up which usually states the exact opposite to the Null Hypothesis, and once again it is particularly useful to state this in terms of the biology you are analysing.

When an alternative hypothesis has no direction, that is, does not imply that one value is larger than another, then it is called a two tailed hypothesis. However, if in the alternative hypothesis some direction of the difference is stated, then the hypothesis is called a one tailed hypothesis. For example, in the t test the Null Hypothesis states $\bar{x}_1 = \bar{x}_2$, a two tailed alternative would be $\bar{x}_1 \neq \bar{x}_2$, a one tailed alternative would state $\bar{x}_1 > \bar{x}_2$ or $\bar{x}_1 < \bar{x}_2$. It is always useful to state the alternative hypothesis in terms of the biology you are investigating.

B. Stage two—Test the Null Hypothesis

Test the Null Hypothesis in two steps:

(i) *Step one—Change your data into standard statistical units*
There is a formula for each test to change your data into the

appropriate standard units and you should use this: for example, for the t test there is a formula for changing your data into t units, for the Chi Square test there is a formula for changing your data into Chi Square units and so on. Basically there are standard units for each test. If you use a computer the computer will do the computation for you and change your data into the appropriate units and give you the answer.

Having applied the correct formula you will now have a number, whose units are in the standard form. We can call this the calculated value and proceed to the next step.

(ii) *Step two—Comparison of the calculated value with standard values from tables*

With your calculated value you can use the standard tables in order to decide whether to accept or reject the Null Hypothesis. To do this you need to be clear about what information the tables tell you. In essence, what the tables give you is the probability of obtaining the calculated value (or a more extreme one) if the Null Hypothesis were true for your data (using your calculated value which is in standard units and which was obtained from your original data).

If you refer to the standard tables you will find that the bulk of the table is made up of various standard values in the appropriate units. As part of the table, usually along the side, there will be labelled the degrees of freedom, and somewhere else the probability levels will be defined (quite often along the top of the table). Look, for example, at the t or Chi Square tables. Now what you have to do is to calculate the degrees of freedom for the test you are using. The degrees of freedom are always set by a formula for the test. For example, for an unpaired t test, in which two sample means are compared, the degrees of freedom are defined as $n_1 + n_2 - 2$ where n_1 is the sample size of sample 1 and n_2 is the sample size of sample 2.

Once you have calculated the degrees of freedom, you can now find the appropriate row or column on your table which relates to this. Go along this row or column until you find where your calculated value falls. Then refer to the table to determine what probability value is associated with that calculated value and note it. Most of the tables do not give exact probabilities but instead give probabilities ranging from about 0.2 to 0.001 in steps. If this is the case and your calculated value falls between two probability values, quote the larger of the two values. This probability value you have now obtained is the probability of obtaining the calculated value (or a more extreme one) if the Null Hypothesis were true. If you have been using a computer program for the statistical analysis, the computer may quote this value for you and so save you looking it up in the statistical tables.

(iii) *Step three—Drawing conclusions*

The convention which has been established is that if the probability for your calculated value is greater than 0.05 (5 per cent), then the Null Hypothesis is accepted since it is considered that there are not sufficient grounds for rejecting it. If however the probability is 0.05 or less, then the Null Hypothesis is rejected since it is considered that it is now reasonably unlikely that the Null Hypothesis is true and there are sufficient grounds for rejecting it. If you think about it for a second, it makes sense. There is only a 5 per cent chance or less that it is true or a 5 in a hundred chance (or less) that it is true with a 5 per cent significance level, so it is relatively unlikely.

If a Null Hypothesis is rejected, then the probability level associated with your calculated value must be quoted and this is the *significance level*. It is a measure of your uncertainty about rejecting the Null Hypothesis since it states the probability of obtaining the calculated value from your data (or a more extreme one) if the Null Hypothesis were true. Sometimes I have seen students quote a 95 per cent significance level. This is a nonsense. If we translate it as it stands, what is being said is that there is a 95 per cent uncertainty about rejecting the Null Hypothesis, so the question would be 'Why reject it then?' In this case, the student probably meant a 5 per cent significance level, but misunderstood how it should be stated.

Having rejected the Null Hypothesis at a particular significance level, you can now accept the alternative hypothesis stated at the beginning of the test. It should be clear to you now why it is so important to recognise, at the outset, what the Null and Alternative Hypotheses are in order to understand what the test is doing in terms of the biology.

There are one or two things worth noting about setting the significance level. Ideally, you should set an acceptable significance level for your data before you start the test. Generally, you could use the guides given above but there are situations where one could argue that high levels of significance, that is, probability values of less than 0.01, are desirable; for example if the outcome of the test determined whether a new drug was to be marketed, or alternatively that it would be very difficult to obtain significance levels of less than 0.1, as is often the case in certain types of behavioural experiments. These factors about significance levels are worth consideration.

6.7   Regression and correlation

Regression and correlation are techniques used where more than two measurements have been made on any one individual in a sample and there are thus more than two variables for each individual. The techniques enable you to define the closeness of the relationships between the variables.

In this chapter, the relationship between just two variables will be considered. (However, multivariate statistics techniques are available for looking at the relationships between more than two variables measured on each individual, but are beyond the scope of this book.)

6.7.1   Correlation

In essence, correlation enables you to express the degree of the linear relationship between two independent variables in terms of a correlation coefficient. The technique is to measure the variables on each individual and then to draw a scatter diagram (a graph showing the distribution of the points). Neither variable has been fixed in any way (that is, values have not been defined) and so each is described as being an independent variable. If there is an indication that the association between the two is linear, the correlation coefficient (r) can be any value from -1 to 0 to $+1$. The nearer the value is to 1 the closer is the linear relationship: a $+$ value indicates a positive relationship, that is, that the variables are directly related (as the value of one variable increases the other increases) and a $-$ value indicates that they are negatively correlated, that is, follow an indirect relationship (where as the value of one variable increases, the other decreases).

The technique can be applied to ratio, interval or ordinal data but in the latter case the Spearman correlation coefficient (r_s) is worked out rather than the correlation coefficient defined by r. Correlation should not be used if one of the variables is time.

The greatest danger in using correlation is that once having identified a linear relationship the temptation is to assume that the relationship is also causal, that is, that one of the variables is responsible for, or causes, the other variable to take on a particular value. Such a deduction should be avoided.

Significance tests can be carried out to test whether the correlation coefficient is significantly different from zero and also to compare correlation coefficients for significant differences. The basic procedure to be followed is as outlined in 6.6.

Usually, straight lines are not drawn through the points on a graph when correlation is used, but if for some reason you wish to, you should use model II regression (see Sokal and Rohlf, 1969), rather than the more commonly used model I regression. It is an easy technique to apply.

6.7.2   Regression

This is a technique for defining and thus drawing the best line through a series of points. Linear regression is used for defining the best straight line and curvilinear regression for the best curve. Here I shall be concerned with the simpler linear regression. In regression, one of the variables, the independent variable, has usually been fixed or controlled by the investigator, for example, a series of concentrations or pH, or temperatures. Some other variable, the dependent variable, is then measured under this condition; for example, it may be transpiration rate, size of muscle contraction or growth. A graph is then drawn on which are placed the points. Various regression formulae are applied in the regression analysis (model I regression is used in this situation) which eventually gives rise to *the* equation which defines the best line through the points. This equation can then be used to construct a line through the points.

In linear regression, the equation is of the form $y = a + bx$. The regression line can be drawn and a significance test can be employed to test whether the line is significantly different from a horizontal line. Significance tests are also available to make comparisons between regression coefficients defined by b in the regression equation (the regression coefficient defines the gradient of the line) and confidence limits can be attached to the line.

Different models of linear regression are available depending on, for example: whether the independent variable is on the x or y axis of the graph; whether there is more than one dependent variable for each value of the independent variable (as occurs when you have data from a sample for each of the treatments for the independent variable); or if both variables are independent variables.

Assumptions are made in the analysis. Some details of the techniques can be obtained from standard textbooks, and Sokal and Rohlf (1969) give a good survey. You may find it interesting that the analysis of variance technique arises out of regression analysis, there is thus a very close relationship between the two. The techniques for curvilinear regression are similar in principle but are beyond the scope of this book.

6.8   A word of warning about applying statistics

I have noticed that once someone begins to 'get the hang of' the statistics tests, there is a temptation to analyse everything in all possible ways, or to continually use the test that the individual is happiest with. This is a serious mistake as it indicates that the biology has not been thought through clearly enough. Use statistics sensibly so that it really does help you in the analysis of the biological data from your investigations and appropriately, so that the techniques used match the design you have. You should, of course, have decided what statistics techniques you would use at the design stage, though sadly many students do find this difficult and consider the statistics at the end of the investigations in spite of warnings not to. Some of the commonest mistakes I have come across are: using non-parametric rather than parametric tests only because they are quicker and easier to use; carrying out multiple t tests when an analysis of variance should have been applied as more than two samples were to be compared; using the Analysis of Variance in many inappropriate situations. There seems to be a constant temptation to use an Analysis of Variance whenever a range of treatments are applied and their effects tested. This happens even though the aim of the investigation is to determine whether there is a trend of response over the range of treatments and not to determine at which points significant differences occur between the treatments. If the aim is to determine a trend, then the levels of the treatments were chosen to give a good range and the values of the treatment levels have no particular significance in themselves except to give a good range. If an analysis of variance is carried out in such a situation, all that will happen is that significant differences will be obtained when the treatment values are sufficiently far apart, and if a graph is not drawn the trend will not be illustrated by just carrying out an ANOVA. If the investigation is really about discovering and identifying trends, as a range of treatments within a factor is changed, then ANOVA between each of the treatments is generally inappropriate, and drawing a graph with points which show the mean and standard errors is much more useful.

Ideally, you should think about how you will analyse your data before you start. If you do this, you can accommodate in your experimental design any restrictions imposed on you by the statistics tests. It may help you to avoid analysing everything in every possible way.

References

Campbell, R. C. (1974). *Statistics for Biologists*, 2nd edn, Cambridge University Press

Daniel, W. W. (1978). *Biostatistics*, Wiley

Denenberg, V. H. (1976). *Statistics & Experimental Design for Behavioural and Biological Researchers*, Halstead

Green, R. H. (1979). *Sampling Design and Statistical Methods for Environmental Biologists*, Wiley

Parker, R. E. (1979). *Introductory Statistics for Biology*, Edward Arnold

Siegel, S. (1956). *Non Parametric Statistics for the Behavioural Sciences*, McGraw-Hill

Sokal, R. R. and Rohlf, F. J. (1969). *Biometry*, W. H. Freeman

Zar, J. H. (1974). *Biostatistical Analysis*, Prentice-Hall

I'M A STUDY ON BUGULA AND I'VE BEEN HERE 9 YEARS!

THERE WON'T BE ANYTHING ON BUGULA HERE!

7

Using the Literature

This chapter will help you to:

Find relevant information in the literature

Recognise what are the main sources of information

Search through the literature

Make best use of your reading

7.1   General points and advice

The scientific literature consists of written reports of scientific investigations and as such will provide you with a very valuable source of information from which to acquire a background knowledge of your project, and to develop ideas relating to: the methods you use in your investigation; the analyses and interpretation of results; and the development of the project through different investigations. It is also important to demonstrate in your report that you have a critical appreciation and understanding of the literature relating to your project. It is therefore necessary to read the literature attentively and critically in order to complete your project satisfactorily.

The amount of literature available can be voluminous. It can, therefore, be a real problem knowing where to start, how much to read, and when to do it. In general terms, you have been encouraged here first to read one or two key sources of information and have discussions with your supervisor to acquaint yourself with the background to the project and the reasons behind the work. You should then start the practical work as soon as possible. Once that is underway, read the literature further to extend your knowledge and understanding. Reading the literature as you do the practical work will keep your reading relevant as the project develops. You will probably find too, that the research papers become more comprehensible, interesting and meaningful as you compare your own work (whether methods, results or ideas) with them. Avoid starting the project with a massive literature review as it will delay the onset of the practical work and probably cause confusion rather than clarification of your own project.

In reading the literature you will inevitably have to compromise to a large extent on what you can do. This is particularly the case when the project is of a limited, short duration and when you have other parts of the course to follow and complete at the same time as the project. You will have to accept that it will be impossible to become fully informed about all aspects of your project. However, in compromising you must make sure that you have read the key up-to-date references and that you have knowledge and understanding of the general history and background of the work leading up to the project.

The advice which follows in this chapter shows you ways of using the literature efficiently (sections 7.3 and 7.4) and starts by giving information on what sources of information are available and how they can be useful (section 7.2).

7.2   Sources of information

Some of the major sources you are likely to find useful are considered below.

MAJOR SOURCES OF INFORMATION

A. *SCIENTIFIC PAPERS*—The publication of papers in scientific journals is the major and fundamental formal channel of communication between scientists where reports of research are presented. When a research worker has reached the stage where he has a sufficient amount of work which is original, reproducible and of general interest in a particular area of biology, he would normally seek to report that work by preparing a research paper for publication in an established scientific journal which reflects the subject matter of the paper. By publishing the work the researcher is basically confident that he has enough evidence to believe that the work reported can stand up to scrutiny by other researchers and therefore that the results have been verified in some way. The paper has to be written in the style and format of the journal and the editor then makes a decision whether the paper is to be published or not, and if so whether changes, additions or revisions are necessary. The editor often seeks the advice of senior researchers in the field in making a decision about publication. There are a large number of different journals dealing with the many different fields and aspects of biology. Some are published by scientific societies or research groups, others by commercial concerns and others by industry or professional institutions.

The quality of the refereeing and scrutinising process varies with the different journals and this has led to the identification of reputable 'core' journals in which the procedure is rigorous, for example, *Nature*. (Your supervisor should be able to help you to recognise what journals are regarded with particular esteem.) On average, it takes about a year for a paper to be published after it has been received. Overall, reporting work in papers is still a fairly rapid and well-controlled way of communicating information in spite of its shortcomings. It would be a mistake though to assume that published work is not open to criticism and that it has to be accepted in its entirety. One of the joys and frustrations of scientific work is that it is open to alternative interpretation. The interpretations in the papers may have to be adjusted because of new developments or because methods may become obsolete—even in the time it takes from receiving the paper to publication. In addition, journals with a less rigorous refereeing procedure may publish work of a poor scientific standard.

B. *REVIEWS*—A review is a lengthy and comprehensive account of the historical development and present position in a particular well-defined area of biology. The reviewer therefore carries out a discussion of work already published and may draw conclusions and take a particular view on the basis of the evidence collated from the work reviewed. A review is normally written by research workers who have made considerable contributions in the area the review is concerned with. With such experience it is expected that the reviewer is able to take a critical, informed view of the published work. A review can be an ideal source from which to acquire a background knowledge of a particular area and an appreciation of the current views and controversies by summarising the existing literature. It is also an invaluable source of relevant references for your project if the review is in a similar subject area. Though the review is expected to be a critical review, do recognise that the reviewer may favour a particular interpretation of the material and may show some bias in the examination of the published material. Some journals, such as *Advances in Genetics* and *Annual Review of Biochemistry*, concentrate on producing review articles only. Such journals are often identified by titles which start in the following ways: *Advances in . . .*; *Annual Review of . . .*; *Progress in . . .*; *Current Topics in . . .*. Some journals also publish both review articles and papers. Depending on the library, some reviews can be treated as books or periodicals/journals and shelved accordingly.

C. *ABSTRACTS AND INDEXES*—These primarily list, in some form or another, the contents of a wide range of journals by subject and are produced monthly or quarterly. The delay between the publication of the journal and its appearance in the abstract or index varies between three and eighteen months. Indexes give lists of journal articles by quoting the title, author and reference. The abstracts give this information and in addition they provide a short summary of the article—an abstract.

The two most useful abstracts and indexes for students on biology projects are: *Biological Abstracts* which covers most areas of biology and *Index Medicus* which lists journal articles from the biomedical literature. Both of these sources are particularly useful for searching for relevant papers through the years. The abstracts are additionally useful in that they provide summaries of the papers, though it is important to recognise that the summary is not a critical review. It does, however, take some time to do an adequate search through the abstracts or indexes. The abstracts and indexes normally have a cumulated index which covers the whole year.

D. *CURRENT AWARENESS PUBLICATIONS*—These publications aim to list articles within a short time of their publication (usually within a week or two). They are generally put together by reproducing the contents pages of the relevant journals published in that week or month. There is not normally a sophisticated scheme of indexing and they are aimed at keeping you as up-to-date as possible. They are not really very suitable for searching back through months and years in the way that indexes and abstracts are. *Current Contents* is the most useful in science and this is subdivided into parts of which the most relevant to biology are: 'Life sciences' and 'Agriculture, biology and environmental sciences'.

E. *GENERAL REFERENCE BOOKS*—These are books which are designed to answer specific factual questions, and are often kept in particular areas in libraries. They include such things as dictionaries, encyclopaedias, handbooks, pharmacopoeias and directories. They are useful for giving specific types of information, for example, the *Merck*

Index is one of the most useful sources of information about chemicals and drugs.

F. *BOOKS*—There is obviously a tremendous variety from very general books on biology to highly specialised books. Generally, the books summarise information in the literature and are therefore very useful for giving an overview of an area of biology to gain some understanding and insight into it. The amount of information you can obtain depends on the degree of specialisation of the book. Books take time to publish, probably at best a year after the final manuscript has been received. The information contained in it and references cited will generally therefore be about two years out of date when a book is published.

G. *MONOGRAPHS*—Monographs are relatively detailed works on a very specific aspect of biology and are often published by Scientific Societies, for example, *Monographs of the Physiological Society*. They are useful in that they are detailed studies of a particular area of biology and can thus supply you with a lot of information and references. However, like reviews, particular points of view may be taken and there is delay in the publication similar to that for books. The distinction with books above is not that obvious and indeed a monograph could be regarded as a book on a very specialised subject.

H. *THESES*—These report the research work carried out by postgraduate students for a higher degree, such as MSc or PhD. A copy of the thesis is deposited in the library of the institution where the degree was done. Theses, even American ones, can be obtained through the inter-library loan system. They can be a very valuable source of information (particularly about practical details) if closely related to your project area, as they often contain more information than the published papers which are produced from them. However, do recognise that the thesis is part of the training for the higher degree so some work may be incorrect or not well-regarded. If the work is generally acceptable to the scientific community it should eventually be published in the scientific journals as research papers. Lists of British theses occur in 'Aslib index to theses' and lists of American and Canadian theses occur in 'Dissertation abstracts international'.

I. *PREVIOUS PROJECT REPORTS IN THE DEPARTMENT*—When projects are continued in a department from year to year, there is a temptation to look at the previous project reports. These can be useful if they provide the basis for your current project and if they are good and cite references well. However, the standards vary tremendously and you do not know what mark was achieved. Do not, therefore, trust all you read in it or use it as a guide of how to write your own. Use the research papers as an example of the correct style and format to apply in writing up.

J. *VERBAL COMMUNICATION*—Scientific meetings are organised for researchers to present their research work to members of the society organising the meeting. The work is therefore up-to-date and on-going. When the papers have been given, they are normally open to discussion by the members. Papers from the meetings may subsequently be published with or without the discussion in the society's journals, for example, The Physiological Society publishes the proceedings in the *Journal of Physiology*. As well as appearing in journals, the proceedings of conferences and meetings etc. are often published in their own right as books and these can also be a valuable source of up-to-date information. Meetings are particularly useful for finding out about current work and for getting to know what views and approaches to various aspects of scientific work in certain areas are in vogue.

In addition to the formal presentation of papers there is of course informal discussion between research scientists both at these meetings and on other non-formal occasions when views and ideas can be exchanged and discussed. It is always worth taking advantage of an opportunity to discuss your work with someone in the same or similar field.

7.3 Finding the information from the literature

The information you need on the biological area of your project may come from any number of sources and much of it will come from the sources outlined in the previous sections. Often the problem can be in finding the right kind of information. The best strategy is to try to find a recent key reference, preferably a review (or some other comprehensive summary of the area of your project) and to use this as a starting point from which to find other references (the recent key references will list other references in its bibliography). Ideally, your supervisor should have provided you with the reference and if he has not you should try to obtain one from him. If you are unable to obtain a reference or if you wish to search for references which are more recent than the paper you have, or if generally you wish to carry out a more rigorous search or obtain information on the same or new area of a project, there are a number of approaches you can take as outlined below.

METHODS OF SEARCHING THE LITERATURE

A. *EXAMINE THE SUBJECT AND AUTHOR INDEX*—It is always worthwhile looking through the subject index of the library to see what books are available in your subject area by referring to the library catalogue. Look generally and specifically; for example, if working on a particular enzyme look up enzymes and see what there is in the library and where the books are located (that is, the classification number) and then look at the books in that subject area. If working on a particular type of enzyme or with a particular organism, look that up too. The books can then give you some general information and are more useful if they are fairly recent. Similarly, if you know the name of an author who is an authority in an area, you can look the name up in the author index in the library catalogue to see what has been published.

B. *BROWSING*—It can often be fruitful to browse through the journals to find some relevant information if you are aware of which journals are likely to publish the material and also if you know an author involved with a particular field. The browsing is made easier by the fact that many journals, though not all, have a periodic index of authors or subjects which you can refer to. If you are not sure what journals to look at, look at the list of available journals within your library, by subject preferably, and see if there are any titles which match your project area. For example, if you are involved with studying animal behaviour, look at 'Animal Behaviour' or 'Applied Animal Ethology' for instance, if they are available. Browsing although undoubtedly useful, can be time-consuming on occasion and fruitless.

C. *REFER TO THE ABSTRACTS AND INDEXES*—Before using these it is important to have clarified exactly what you want to find out, which means knowing the subject of your project very clearly. This, of course, should be evident from the title and aims of the project. From this, you should be able to identify the subject area of your project and key words associated with it. For example, a key word may be the name or group of an organism, such as insects or *Pieris brassica*; it may be the variables you are measuring in your investigation, such as photosynthetic rate, or the general area of investigation, such as respiration, biodegradation, etc. Try to have more than one key word for flexibility.

Always start with the most recent abstracts and indexes and work backwards in time. It is also useful to use the yearly cumulated index or abstract to enable you to look at one year in one go.

BIOLOGICAL ABSTRACTS—there are six main sections which are listed in the inside of the front cover of each volume of the index where some instructions are given on what each section is and how to use it. The six sections are

Subject index—Which lists key words which identify a particular aspect of the subject, such as slugs, molluscicides, sleep etc.
Biosystematic index—Indexes papers according to their biological taxonomic groups, for example, Angiosperms, Dicotyledons, and families within the dicotyledons.
Generic index—The species are listed in alphabetical order of generic names.
Concept index—Indexes papers according to broad subject areas (concepts) in alphabetical order, followed by subheadings; for example, Genetics and cytogenetics is the concept area, genetics and cytogenetics—animal, and genetics and cytogenetics—general are subheadings within it.

Author index—Authors are listed in alphabetical order.
Abstracts section—Lists by number, within broad subject areas, abstracts of papers. The title, authors and full references are given together with the abstract.

All sections, except the abstracts section, list the abstract numbers relevant to the section referred to; for example, there may be a number of papers by one author and each is referenced to by a number for an abstract. There may be a number of papers within a subject area and each is listed by an abstract number. The usual way to begin to search for relevant papers in the abstracts is to identify key words. For example, suppose you wanted to test the effects of different osmotic pressures on fruit cells. Key words could be 'osmotic pressure', 'fruit', and an additional possibility would be 'cells'. You can then look up the key word in the subject index which will list the titles of papers with that word in the title of the article. The title starts at the symbol /. Beside each title is a number which refers to the abstract in the Abstracts section. On looking up that reference number you will find the abstract to the paper and the full reference. If the abstract looks relevant you can take a note of the paper and look it up in full, or put in a request for it through inter-library loan. The abstracts themselves are arranged in broad subject groups. Key words are derived from the title. The spelling will be American. It is always worthwhile having a number of key words and using as many synonyms as possible to take account of the variations in usage of the key word. Other sections of the Index can of course be additionally referred to.

INDEX MEDICUS—This index uses Medical Subject Headings (MeSH) for indexing rather than key words. There are two sections: the subject and author index. The author index is arranged in alphabetical order of the author's name and gives the title of the paper and reference source for the particular author. The subject index is arranged in alphabetical order of Medical Subject headings. Within each of these are listed the title of the article, the author and the reference. Subject headings for the index are listed separately in 'Medical Subject Headings' and ideally you should look this up first. If the subject term you plan to use is a used MeSH term it will appear in large letters, if not it will be in small letters and you will be referred to the suitable MeSH term. For example, suppose you were interested in the anticoagulant phenprocoumon; when this is looked up in 'MeSH' it appears in small letters and you are referred to the term COUMARINS, which is the suitable MeSH term.

D. *EXAMINE CURRENT CONTENTS*—The contents pages of current journals are listed and the index of the journals covered in any edition are shown at the beginning of the edition. You could, therefore, refer to the contents pages of journals you were interested in. There is also, in the current contents, an author index in which authors of articles from the contents page of the journals are listed in alphabetical order. If you refer to the authors a number is given which refers to the current contents page on which the journal contents begin. As well as these, there is a title word index in which key words from titles are listed in alphabetical order. For each key word listed, the page reference for the current contents and the page number for the journal is given which can be referred to to identify the paper within the 'Current Contents'.

E. *COMPUTER SEARCHES*—Computer searches for references may be carried out by your library and if this is the case one of the library staff will carry out the search for you. The library will have facilities for access to certain bibliographic databases which store information on published material. Using key words, relevant literature within certain time periods can be listed with their source. It is, therefore, important once again to identify key words carefully. One of the refinements of the system is that it can look at combinations of key words and identify papers which only have a certain combination; for example, a general key word could be 'toxicology'. A large number of references could be identified. If the combined key words were 'toxicology' and 'malathion', the number would be reduced and if the key words were 'toxicology', 'malathion' and 'insects', the number of papers identified would be fewer, since there would be fewer papers with all these key words occurring together.

7.4   Some advice in using the literature

PRACTICAL ADVICE WHEN USING THE LITERATURE

A. *KNOW THE LIBRARY*—You may have been lucky enough to have been given a proper introduction to the library by the librarians so that you know what is available. If not you will have to find out what services your library supplied; in my experience, library staff are friendly and willing to help if you have difficulties and many have prepared material to help you. You need to establish what journal collection is available, what abstracts and indexes are stocked, where current periodicals are kept, the location of the subject and author index, what inter-library loans and computer search facilities are available to you and what photocopying services are present. Generally, you will not be allowed to take journals out of the library and so will have to read the papers in the library or obtain a photocopy to take home to read. It is possible that some libraries loan journals overnight.

B. *CONSIDER JOINING ANOTHER LIBRARY*—It may be possible to obtain a reader's ticket for another library which may offer better facilities than yours. For example, the local University is likely to stock a good range of journals and you may be able to use it. The best thing to do is to go to your librarian and/or the library you wish to join and ask about the possibility of joining.

C. *TRY TO IDENTIFY KEY REFERENCES OR AUTHORS EARLY ON*—Try to find out, ideally from your supervisor, a key reference or author. This can then act as a good starting point for your examination of the literature. If your supervisor is unable to help, you will have to use any of the search methods given in section 7.3.

D. *AIM TO KEEP YOUR READING RELEVANT*—At all times examine closely what direction your project is going in and read according to the way the project is developing. For example, you might start by looking at the effects of heavy metals on mitochondrial respiration but end up looking at methods of isolating mitochondria for producing reliable preparations. The reading in each case has a different direction—formerly respiration and heavy metals and latterly isolation methods.

E. *START A CARD INDEX FOR REFERENCES*—Right from the start of your reading it is essential to keep a systematic, accurate and careful record of what you read. An ideal way to do this is to keep a card index with one card for each reference. On the card will be details of the author, the title and the full reference of the article or book. These cards can then be sorted in different ways, such as subject index or alphabetical order of author. Do keep accurate records from the start and then the cards can be used directly for preparing your bibliography in the report. It is often worthwhile, too, making a note of something that interests you in the article on the card. It is important to keep an accurate record of your references as it is extremely unlikely that you will be able to remember the source.

F. *SEPARATE OUT WHAT HAS BEEN READ FROM WHAT YOU PLAN TO READ*—Keep a clear set of cards you plan to read from those you have read, and then when you have time you can go to the library and go through the various cards for references for articles you plan to read.

G. *KEEP RECORDS OF THE INFORMATION YOU HAVE READ*—Some information on what an article says can be kept on the card, or if you need to make a lot of notes keep them on file paper. You might also find it useful to begin to separate out on different pieces of paper points you wish to develop in the various sections of the report, for example, one piece of paper on points and ideas for the Discussion, another for Methods, etc.

H. *DO READ AS YOU ARE DOING THE PRACTICAL WORK*—The practical part of the project is an important part of your project and you will find that often details in the literature may help you with developing any of the various parts of a project, for example, planning and designing investigations, sorting out methods, interpreting and analysing results. Furthermore, as noted earlier, the direction a project takes depends on what is happening with the practical work.

I. *TRY TO PUT TIME ASIDE TO SEARCH THE LITERATURE*—It is worth going through the abstracts for the last one or two years, or to examine the years between your most recent references and the current year, since one good recent paper can provide you with a list of references. Do, however, recognise that it does take time to do a reasonable search, so you need to leave several hours for it. Also, use the other sources and approaches as outlined in the previous sections to search the literature.

J. *WORK FROM THE GENERAL TO THE SPECIFIC*—In order to acquaint yourself with a subject, especially if it is not very familiar to you, it is best to start from a general source, such as a good textbook or review, and then when you have a grasp of general aspects of the subject, look at the detailed papers.

K. *RECOGNISE YOU MAY HAVE TO READ A PAPER A NUMBER OF TIMES TO GRASP IT*—There is a lot of condensed and summarised information in a scientific paper and it is written for a well-informed group of people. It can, therefore, take a lot of effort to understand a paper, especially when you are starting out on a piece of work. The more you read it, hopefully the more you will understand it. If it is a key paper, read it on different occasions and fairly regularly. You might find that as you grasp the subject of your project more thoroughly, you see new meanings and make new interpretations on the same paper.

L. *AIM TO LOOK THROUGH CURRENT JOURNALS REGULARLY*—Try to put time aside to browse through the current journals and current contents once a week to see if there are any relevant papers. Remember to take your card index with you.

M. *TRY TO BE CRITICAL AS YOU READ*—Initially you will find that in reading the papers your major task is to become familiar with the literature by understanding the papers. With experience, you will acquire knowledge and should try to take a more critical approach in your reading. Chapters 5 and 8 offer some suggestions on how to do this.

N. *LOOK AT THE RESULTS PRESENTED CAREFULLY*—When examining the results presented in a paper, make sure that they show what is claimed, that they have been analysed appropriately and that the interpretations placed on them are reasonable (see chapter 5).

O. *TRY TO APPRECIATE THE STYLE OF PRESENTATION*—Your report will have to be in the form of a published paper, so recognise what the style and format is. Within the paper there will also be a critical view taken of the work which might help you in carrying out a critical review of the literature and your own results. Notice too how other references and authors are referred to and the style of presenting results.

P. *DO NOT RELY ON THE ABSTRACT AS YOUR ONLY SOURCE OF INFORMATION*—Once you have the abstract of a paper it can be tempting not to go further and not to look up the full paper. This is poor practice as the abstract is an uncritical summary and you need to read a paper fully to gain any real appreciation of it.

Q. *ATTEND MEETINGS WHEN YOU HAVE THE OPPORTUNITY*—If there is a scientific meeting of any kind—departmental, national or whatever—it is worthwhile going to see how original research work is presented and discussed, and of course if it is in the area of your project, it will also be relevant and hopefully useful.

IRREFUTABLE PROOF!

8

Writing the Project Report

This chapter will help you to:

Write a project report in the accepted scientific style

Develop coherence in your report

Demonstrate your knowledge of the literature

Write a critical appraisal of the work

Recognise some of the practical problems in writing the report

Appreciate the importance of writing a good report

8.1   The importance of writing a good project report

Students often underestimate the value of writing a good report of the project, so it is important that *you do* recognise the value. The written report can have the following effects.

THE KINDS OF EFFECTS THE PROJECT REPORT CAN HAVE

A. *SIGNIFICANTLY AFFECT THE FINAL MARK OF THE PROJECT*—A good project (in terms of the scientific quality of the investigations) can be pulled down by a poor write-up and a poor project can be lifted. The report can have such an effect that it could alter (in either direction) the final class (in a degree) or grade (in a BTEC course) awarded to the project. (Remember the report might be all or most of what the examiner has to go on in making his assessment.)

B. *BE OF VALUE EDUCATIONALLY*—Not only will it give you experience at writing in a scientific way, it will also give you a clearer understanding and greater insight into what you have been doing in the project. You will need to be disciplined in your thinking to write the report carefully and thoroughly.

Do not, therefore, take the view that the presentation of the project does not really matter as the quality of the work will shine through— it will not—and *do* make every effort to produce a good report and recognise that it will take time to do this. For a project report to be good it should show certain characteristics.

CHARACTERISTICS OF A GOOD REPORT

A. *COHERENCE*

B. *GOOD PRESENTATION*—In all its aspects

C. *A THOROUGH RELEVANT BACKGROUND TO THE PROJECT WITH CLEARLY STATED AIMS*

D. *A CAREFUL CRITICAL ANALYSIS, INTERPRETATION AND DISCUSSION OF THE RESULTS WITH PERTINENT REFERENCE TO THE LITERATURE*

The text which follows (sections 8.3 to 8.9 inclusive) offers suggestions and general advice on how to achieve these. It is also very useful to read the papers in the literature with care. They are examples of accepted scientific style (though the quality varies) and illustrate many of the points made in this chapter. They will also be a source of ideas for material to be included in each of the sections of the report.

> *Example*
>
> The Introductions of the articles give the backgrounds, the Discussion sections show ways of explaining results, and each of these sections carries out some critical review of the literature and the results of the investigations being reported. They thus may give examples of the types of criticism which can be made in certain areas of work. The papers also provide a valuable source of references you might wish to follow up.

8.2   General practical points to be aware of before you start

There are a number of general aspects to writing a project report which you should be aware of before you start. Hopefully you will then take account of them when planning how to use your time profitably.

PRACTICAL ASPECTS OF PREPARING A PROJECT REPORT

A. *MAKE ENOUGH TIME TO WRITE THE REPORT*—Do not leave the writing up to the last minute. It is difficult to give the time it might take, as individuals vary so much. However, suppose that a project is to be 3000 words long (or at least the Introduction and Discussion of that length). Consider how long it might take you to write up a single (3–6 hour) extended practical and use this as a guide. As well as doing the writing you will also have to leave time to:

  (i)   prepare graphs and tables, diagrams and photographs, so that they are well presented;
 (ii)   get your supervisor to read a draft copy and then make corrections;
(iii)   prepare the bibliography so that it is correct and full;
 (vi)   have it typed;
 (v)   check the typed copy for errors and then have it corrected;
(vi)   get it bound if necessary.

With all things considered it could take you a similar number of hours to prepare the project report as the time allocated for the project practical work.

B. *WRITE THE REPORT ACCORDING TO THE FORMAT SPECIFIED BY YOUR COURSE*—Projects are normally written up in a similar style to research papers and by reading these, which you should be doing during your project, you will come to appreciate the style. In general, there will be a limit to the length of the project report and it will have to be divided into a number of subsections. (Commonly these sections include: Introduction/Literature survey; Methods; Results; Discussion;

Bibliography; Summary; Acknowledgements.) It is customary to give the title on the first page with the name of the author, the award for which the report is part and the year of submission. It is useful then to give a list of contents with a page reference and then the summary.

An appendix can be included, if you feel it is necessary to present material which, though relevant to the subject, is particularly detailed or lengthy and which if included in the main text would distract the reader from the main argument.

C. *WRITE WITH YOUR READER IN MIND*—Think about what difficulties he might have in understanding anything you present and aim to minimise those difficulties while recognising that the reader has some knowledge of the subject.

D. *ENSURE THE WORK IS WELL PRESENTED*—Some detailed advice is given in preceding sections on the style of presentation of particular sections. However, in general write in good English, in full sentences, vary their length, avoid using the same phrases over and over again, and keep within the limit of the number of words recommended. Take care over the punctuation, grammar and spelling. Do make sure you correct typing errors in the final report. Take account of writing style. It is often easier to follow relatively short sentences rather than long ones with a number of clauses in. Your reader will find it easier to follow a text if you state the point you wish to make and then develop it, rather than use a style in which you give a lot of information first, leading up to the point of it. Try not to ramble, which can be avoided by thinking out clearly what you want to say and by planning the paragraphs and their organisation to one another. Use headings, lists and diagrams whenever you feel they make a point more clearly, and try to ensure that the project holds together as a coherent piece of work with a beginning, a middle and an end.

E. *MAKE SURE A DRAFT COPY OF THE REPORT IS READ*—Your supervisor, or another lecturer with knowledge of the subject, could read critically a draft copy of the report. Take careful note of any advice given in order to make any necessary adjustments to the report. This cannot be overemphasised. Read the draft copy critically for yourself before presenting it to the supervisor for comments.

F. *KEEP A COPY OF YOUR WORK*—The loss of work could cause you many problems, so keep a copy if you can.

G. *DO NOT FORGET ABOUT THE WORK ONCE THE REPORT IS HANDED IN*—You may have a *viva voce* in which it is likely that the examiners will concentrate on, or start by asking questions on, the project. Be prepared to discuss it in an oral exam, therefore.

8.3   Maintaining coherence in the project report

The project report should be presented as a coherent piece of work, held together by a central theme, focussed by the title and overall aims of the project. The report should not, therefore, be written up as a series of separate practicals or as a diary of events. Instead you have to pull all your work together and present it as a single, major study. Normally, therefore, the project report is divided into a number of clearly defined, related sections as given below.

THE SECTIONS OF A PROJECT REPORT

A. *THE TITLE*—The overall aim or subject of the project.

B. *THE INTRODUCTION/LITERATURE SURVEY*—A background to the project with the aims clearly stated.

C. *METHODS (AND MATERIALS)*—A full unambiguous description of the methods used.

D. *RESULTS*—A summary of results clearly presented and analysed.

E. *DISCUSSION*—An explanation, discussion, debate of the results within the context of the aim of the project and other relevant studies or investigations.

F. *BIBLIOGRAPHY*—A list of references cited in any part of the report.

G. *SUMMARY*—A brief summary of what was done, obtained and discussed.

H. *ACKNOWLEDGEMENTS*—A general statement acknowledging help given by individuals, institutions, departments etc.

Your aim in writing the report is to ensure that there is some coherence between the sections and a natural flow of information and ideas from the beginning to the end. Many students experience a number of difficulties in deciding what exactly to put into each section—particularly the Introduction and Discussion. To help you to decide, it is useful to consider the project report like a mystery story. The mystery is identified early on and then the events which occur as the mystery unfolds are described. Each of the sections in the report has a particular part to play: the scene of the mystery is set out in the Introduction; the methods employed to solve the mystery are described in the Methods section; the outcome of the investigations into the mystery presented in the Results section; and in the Discussion an analysis is given of how far or how much the mystery has been solved. Working like this will help you to develop a theme and coherence between the sections. It should also help you to decide what to put into the Introduction or save for the Discussion, as you will not want to 'give the game away too early' (that is, explain your results in the Introduction).

8.4   When and where to start

Planning

Once you have gained an overall impression of what you have to do, the difficulty can be in knowing where to start. Start planning from the *results*. The results can play a central role in determining the direction taken by the other sections and have the effect of pulling the report together by giving it a clear focus. The direction each section in the report will take in consequence is shown below.

THE WAY IN WHICH SECTIONS IN THE REPORT ARE INFLUENCED BY THE RESULTS OBTAINED

A. *THE METHOD*—Is an account of what you did in order to obtain the RESULTS you got.

B. *THE AIM*—Is a precise statement of what you planned to do in the methods to achieve all the RESULTS obtained.

C. *THE INTRODUCTION/LITERATURE SURVEY*—Gives enough general background and specific information to explain the purpose for and reason behind the aim, which reflects the RESULTS.

D. *THE DISCUSSION*—Expands out from the RESULTS, explaining what they might mean, whether they were expected, how the methods may have affected them, and so on.

The following flow diagram also illustrates the point, showing you the starting point at the RESULTS, how you should proceed and the direction of information as it appears in the final report.

Flow diagram showing the direction in which a report is planned and finally read

Directions for planning the report

Direction of flow of information as it appears in the final report

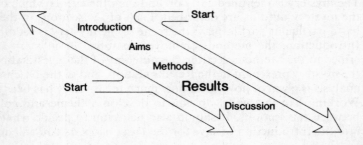

The scheme of work presented provides you with a way of planning your report and should help you maintain some coherence between the sections, avoiding the pitfall of writing sections which seem to be divorced from one another. There is a particular danger of this happening if your project departed in a significant way from your or your supervisor's original plans. It would make little sense, for example, reviewing the background in the Introducion for a project which though planned, was not actually carried out, since the results presented and the Discussion would be on a different theme. Coherence would be seriously lost. Adjust the report, therefore, to what was done.

Example

A student was examining the effects of nematicides on a particular species of nematode. The species became unavailable and he needed to obtain results quickly so decided to also work on an insect plant-pest species. The overall aims were modified to examine the effect of pesticides on the feeding behaviour of two different types of pests. The Introduction would now look at general aspects of plant pests and their control, and more specifically in relation to nematodes and insects. Previously, the Introduction would have concentrated on the nematodes.

One word of warning, however: always check with your supervisor that this approach is acceptable. It might be possible that the supervisor was primarily interested in seeing how you solved the actual problem which was set.

Start then by gathering together the summaries of all your results. Make sure you include all your results, that is, both the negative and the positive ones, those that are clear-cut, those that are not, the results of preliminary investigations etc. List all the results in the form of clear unambiguous summaries which are concise and precise. They will be in the following form. These are basically conclusions from the results analysis.

Example

No separation of the blood proteins extracted from rat blood occurred. The mean feeding bout of 6 week *Pieris brassica* larvae on cabbage was found to be 2 ± 0.6 ($\pm$ standard error, $n = 10$) min. The peak absorbance of the water extracted pigment from the alga was 400 nm.

Organise them into related groups if it is possible, giving each group a clear descriptive heading. You might find in the end that the results are organised best in the order in which they were obtained or as preliminary and main experiments. Alternatively, there may be different approaches or investigations which categorise them or they might be grouped as major and incidental results. Use these to guide the major sections of the report.

Plan the rest of the project report by writing a sentence or two which summarises what you will put in each of the major sections, using your results to guide you.

Example

The effect of feeding stimulants on the feeding behaviour of a particular species of caterpillar was investigated. The results overall showed that the feeding stimulants had an effect on the feeding bouts. Differences between the stimulants used were observed.

Directions planned for the other sections

INTRODUCTION—Examine some of the general background to feeding behaviour in insect larvae and the role of phago (feeding) stimulants. The reasons for using the particular stimulants in the present study to be identified.

AIM—State that the effect of certain named feeding stimulants on various aspects of the feeding bout of the caterpillars was investigated.

METHODS—Describe the methods used to obtain the results on the feeding bouts.

DISCUSSION—Examine possible reasons why the stimulants had the observed effects, including why the effects were different between the stimulants (a detailed examination of the control of feeding behaviour may be necessary).

Starting the writing

With the overall plan ready, you can start the writing. It is probably best to start by writing either the Methods or the Results. The Methods almost 'write themselves' as they are an account of what you did, and the Results should be so familiar to you by now that you are clear of what needs to be described. The advantage of starting with these sections is that two sections get written relatively quickly and can create a sense of 'well being' and confidence as quite a substantial part of the report is accomplished. In addition it can get you into the habit of writing. When you finish these sections, go on and write the Introduction and Discussion, choosing an order which suits you best.

When you do start writing it is useful to keep separate pieces of paper for each section for notes on writing that section. (As you develop ideas you can write them on the relevant piece of paper.) When you come to write the sections you will then have a few ideas of what to include already written down. It might also help you to avoid repeating the same information in different parts of the report.

Separate out too the record cards for references you are quoting. If you did not keep reference cards, make a list or start the card index now. You will find that writing the Bibliography at the end can then be made much easier and less time consuming.

8.5   Writing the methods section

The Methods should be an account of what you did in the project, written in the past tense, in the objective rather than subjective style, that is, 'The absorption was measured' rather than 'Measure the absorption' or 'I measured the absorption.' Draw together the relevant information for the method in a coherent way, rather than giving a week by week guide of each practical. Ensure that you summarise what you did so that there is enough detail for someone to either repeat the experiment or compare his method with yours, but do not give unnecessary detail like 'I labelled three test tubes.'

If when you are planning the method, the material you wish to include seems to fall into natural sections then, as a general guide, let it, and have a number of subsections within the method. For example, you may wish to have separate headings for 'materials (with source)', 'rearing conditions', or 'preliminary experiments to develop a method for . . .' etc.

When writing the method you should make sure that you give clear information on the following, when it is relevant to your project.

INFORMATION TO BE INCLUDED IN THE METHODS SECTION

A. *DETAILS OF THE POPULATION OF ORGANISMS INVESTIGATED*—For example, sex, weight, age, strain etc.

B. *THE SOURCE OF THE BIOLOGICAL MATERIAL ON WHICH YOU WORKED*—For example, a supplier or from the wild, and which supplier or from where exactly was the material collected, and how?

C. *THE CONDITIONS THE ORGANISMS WERE KEPT UNDER*—Both after collection and before being investigated. (Include the environmental conditions, details of nutrients etc.)

D. *DETAILS OF THE METHODS YOU USED TO OBTAIN THE MATERIAL ON WHICH MEASUREMENTS WERE MADE*—For example, the dissection procedure to obtain a muscle, the extraction method for a particular biochemical, the fractionation procedure to obtain an organelle like a motochondrion etc.

E. *THE PRECISE VARIABLES MEASURED*—For example, the length of the coleoptile from its attachment at the seed to the tip, the peak absorbance, maximum width etc.

F. *THE PRECISE DETAILS OF THE INVESTIGATION FOR BOTH PREPARING THE MATERIAL AND FOR MAKING ANY MEASUREMENTS OR OBSERVATIONS*—These details must be those you used and may differ from what you planned, for example, the temperature may be 21 ±1°C rather than 21°C. Include details like the weight or sex of the animals, the age of the organisms and dilutants; consider whether the time of the day or year of the investigation is relevant, how was an animal placed in a test box, if a solution was constantly stirred, the time microorganisms were left to grow before being examined etc.

G. *THE EXACT TECHNIQUE USED FOR MAKING MEASUREMENTS*—If it is a standard technique you may be able to refer to the original author and give the reference in the bibliography.

H. *THE MAKE AND MODEL NUMBER OR IDENTIFICATION*—For any instrument used.

I. *PRECAUTIONS*—taken in preparing and maintaining material or in making measurements, such as the machine was left for 10 minutes to warm up, the animals were left for half an hour in the box to acclimatise, the solutions were kept on ice at all stages of the preparation, plates were prepared in a flow chamber, standard sterilisation precautions were used, etc.

J. *LEVELS OF ACCURACY*

K. *THE TYPE AND SIZE OF SAMPLE*—For example, matched or unmatched and the method of obtaining that sample. The number of replicates.

L. *A BRIEF STATEMENT ON HOW THE RESULTS WERE ANALYSED*—For example, a 't' test was carried out to test whether there was a significant difference in the mean length of larvae kept in blue or red light. Identify *exactly* what variables were analysed.

It surprises me how often students write muddled methods—perhaps because they are 'too close' to the work. A useful way to ensure clarity is to give a step by step guide to the methods according to the steps actually carried out (identified from the list in section 3.2) and the techniques used (section 3.3). As you describe the steps and techniques used, you can refer to the previous list to make sure you have included enough relevant information.

8.6   Writing the results section

In this section your main aim is to present your results to the reader as clearly and precisely as possible. There should be enough information to make it an honest representation of what you obtained and to confirm that the results really do show what you claim. You will, therefore, have to decide which method is the most suitable for presenting your results. This could be written descriptions of what occurred, tables, graphs, diagrams, drawings, photographs or summary statements. In addition, you will be expected to write a text in the results section which identifies for your reader what the results show, which refers the reader directly to the various summaries. In doing this you should make sure *all* summaries are referred to. Students rarely write a text in their Results section and this is often criticised by examiners as being a weakness. A quick look at the papers in the Literature will confirm for you that it is 'the norm' to write a text and will also show you the style of that text.

Quite often results are spoiled by sloppy or poor presentation which could be avoided by some thought and organisation. The following points are suggestions about how to present results and take account of the common mistakes made by students. But first, make sure you have clear precise summaries of your results from your analyses.

FACTORS TO TAKE ACCOUNT OF IN WRITING THE RESULTS

A. *DECIDE ON THE BEST METHOD TO PRESENT YOUR RESULTS*—You should choose a method which illustrates most clearly what you want to show; for example, if you want to show that there is a linear relationship between two independent variables, draw a graph rather than give a table and give the correlation coefficient on the graph.

B. *DO NOT INCLUDE ALL RESULTS IN ALL POSSIBLE WAYS*—All your results do not have to be included in all possible ways. For instance, there is no point presenting exactly the same data as a graph and as a table: if you have worked on taking measurements from a sample there is often little value in giving every measurement—a mean±the standard error with the sample size in brackets would convey the information; if you have a number of chart recordings (of, for example, the effects of drugs on the spontaneous activity of the small intestine) you need only select the trace which illustrates the effect most clearly and should point out that this represents a typical example;

similarly if you are working with the electron microscope you do not have to show a photomicrograph of every preparation—if you do feel the raw data should be included it is often better placed in an appendix.

C. *ENSURE THAT EACH DIAGRAM, TABLE ETC. HAS—*
  (i) A reference, such as Fig. 6.
 (ii) A heading which states exactly what is being shown. For example, standard curve is not enough because the question arises, standard curve of what?
(iii) A key to all the symbols, colours etc.
(iv) Measurement units are given and given correctly.
One way of testing whether the labelling is adequate is to see if the table etc. is understandable on its own without further explanation—it should be. You could ask a friend to check to see if he can understand what the graph etc. is about.

D. *DECIDE WHETHER THE RESULTS MIGHT BE MORE CLEARLY PRESENTED IN SECTIONS*—For example, one section on 'the effects of temperature on enzyme activity', another could be on 'the effects of pH'. Give each section a title which indicates what is in the section. Avoid vague titles like experiment I. It does not really show what the section is about and is therefore of limited value as a heading. If you have divided the methods into sections it is often helpful to have similar sections in the results, where relevant.

E. *PLAN THE PRESENTATION OF GRAPHS AND TABLES CAREFULLY*—Consider using a variety of symbols on a graph or colours (but remember to adjust a photocopy when there are a number of graphs on the axis). Think about whether you wish to make comparisons between sets of results and, if so, it might be better to place these on the same page. For example, you may wish to place a series of graphs under one another, superimpose them or put them all on the same graph. If comparisons are to be made, keep to the same scale for graphs or the same labelling for rows and columns of a table.

F. *CONSIDER A SUMMARY TABLE FOR STATISTICAL DATA*—If you have carried out a number of statistical tests, it is useful to make a table which summarises what was compared with what and the outcome of the test (significant or non significant and the level of significance). This then acts as an easy reference for your reader to see what was found.

G. *STATE THE OUTCOME OF STATISTICAL TESTS PRECISELY*—When presenting the results of a statistical analysis, ensure that you are extremely precise in your statements. There is a tendency to write a statement like 'The "t" test showed a significant difference' but the question arises a significant difference between what? So identify what was being tested and the outcome clearly. This often means identifying which variable was analysed and the conditions of the investigation; for example, using an unpaired 't' test a significant difference was found between the mean time period of a feeding bout for 5-day-old *Pieris brassica* larvae fed on cabbage leaves compared with larvae fed on lettuce leaves (significance level 0.01, $t=3.2$). The reader then knows what was analysed, the outcome and the test used.

H. *WRITE A TEXT FOR YOUR RESULTS SECTION*—This text should be a description of what was found and *not*, as a general principle, a discussion. You should therefore *describe* what the results show and refer your reader to the diagrams etc. as necessary, ensuring that you do refer to all the summaries. For example, the text would be of the following style: '15 out of the sample of 20 animals fed and the analysis therefore is on the 15 animals. The values for the consumption rates and the absolute assimilation rates are shown in table 2 and the relationship of each with temperature is shown in graph 1. From the graph it can be seen that the relationship of each of the variables with temperature is linear. A linear regression analysis was carried out to fit the best straight line and the regression equations are given on the graph. Regression was significant at the 5 per cent level of significance. These results suggest that consumption rates and assimilation rates are temperature dependent.'

I. *BE PRECISE IN THE TEXT*—In writing the text always aim to be as precise as you possibly can and avoid being woolly, rambling or vague. For example, rather than write 'there is an increase' write 'there is a mean increase of 5 per cent . . .', rather than 'generally . . .' write 'in 90 per cent of the cases investigated the plants were unhealthy. An unhealthy plant was identified as one in which . . .' etc. Sometimes, in doing this you may find you have to go back to your original results and group them in different ways.

J. *BE HONEST*—If you did not get a response, if it was not clear-cut or if responses were rarely found, say so. Do not claim too much from the results obtained.

K. *AVOID CONTRADICTING YOURSELF IN THE RESULTS*—For example, there is a common tendency to present the results of a significance test which shows no significant difference but then to claim that there is a difference really! (only because the experimenter did not believe or understand the outcome of the test).

8.7 Writing the introduction and discussion

The Introduction and Discussion are written essentially in essay style, in which points are made and developed in paragraphs and the paragraphs are arranged in an order to produce a natural flow and development of ideas. The overall theme of the Introduction is that it should provide the reader with a background and justification of the aims of the project and state those aims unambiguously. The Discussion, on the other hand, should explore various explanations of the results within the context of both the current investigations and those carried out by other researchers. Relevant points relating to these themes have therefore to be identified and included in the appropriate section. In addition, it is expected that in these sections in particular you should demonstrate a knowledge and understanding of the literature and make a critical assessment of your own and others' work.

A useful way to write these sections is to start by deciding what relevant general points and ideas are to be included and developed in each. After that a plan can be made to ensure a flow of ideas, then write in good essay style, taking account of past advice and experience, and demonstrate a knowledge of the literature and take a critical view.

The following parts of this section offer help on what to include in each section (8.7.1 and 8.7.2). Section 8.8 suggests a style of writing you could use to demonstrate a knowledge of the literature, and section 8.9 offers techniques you could apply in the writing to show your ability to review work critically.

8.7.1   Deciding what points to include in the introduction

Place in front of you the title and overall aim of the project and the specific aims of the component investigations (adjusted if necessary from the original). You will now have to plan an Introduction which gives a background and justification to those aims and which states the aims clearly and unambiguously (see chapter 2). To give you some ideas and to help you to decide what to include, examine the points outlined in the list below and consider including these in the Introduction.

AREAS WORTH CONSIDERING FOR INCLUSION IN THE INTRODUCTION

A. *THE GENERAL AREA OF BIOLOGY YOU ARE CONCERNED WITH*—(This will also help you to decide on the theme of the Introduction) For example, in a project you may be determining the characteristics of a named enzyme, but think about what you are really concerned with—is it in *developing* assay techniques to allow you to determine the characteristics, or is it more generally about identifying the enzyme characteristics, once extracted, using routine extraction procedures. In the former case the area of primary concern is techniques, in the latter the enzyme itself specifically. The emphasis in the introduction should reflect this.

B. *BACKGROUND INFORMATION*—What is known about the work relating to your project and to the general area of biology you are concerned with? What background information will the reader need to be aware of to appreciate what you are doing?

C. *PROBLEMS/DIFFICULTIES*—What problems are there in your field of study, why are they problems and what attempts have been made to overcome them? Were they successful and what other details are known? What problems have other workers in similar fields had and will your work help to overcome these difficulties? If so, in what way?

D. *GAPS*—What are the gaps in the general understanding or knowledge of the subject? Has this led to your current investigations? How much is your work likely to fill those gaps?

E. *REASONS FOR DOING YOUR PROJECT*—What led to your particular project and investigations? For example, is it a development, extension, repetition of previous work? If so, why was that necessary or useful? Is it something new—a different approach, another method, a different species or type of organism? Will it lead to the clarification of a problem, create a better or more complete understanding? How is your work unique? Are you testing a specific hypothesis?

F. *POSSIBLE OUTCOMES OF THE WORK*—What do you hope to achieve in the investigations and how can that help our understanding of the system being studied?

G. *JUSTIFICATION OF YOUR APPROACH OR METHOD*—Is there some controversy surrounding what is the best method, or their limitations? Is there a particular reason why you are taking a particular approach or using a technique, method, or instrument?

H. *IMPORTANCE*—In what way is your work likely to be important?

Having surveyed this list, *write down* the points you wish to develop in the Introduction, ensuring that each is relevant to the general approach of the project and to the specific aims of the investigations. *Organise* the paragraphs so that there is a natural flow and development of ideas. A useful way to do this is to start from the more general aspects of the project—giving the background and introducing your reader to the overall area of the project. Then develop the points and ideas to provide a justification for the aims of the project. These are finally stated at the end of the Introduction very clearly and specifically. Overall then, the Introduction 'sets the scene' for the project.

Example

Title of project 'Biodegradation of hay'

The main reasons for the project were twofold; first because little work had been done on good hay compared with poor, and second the work that had been done indicated that there were risks in developing certain allergies or infections from the good hay as well as the poor hay.

An approach which could have been taken in the Introduction would be to give first a very general account of what occurs in the biodegradation of hay when it is being stored by farmers, its importance (good and bad) and a consideration of what is regarded as good and poor hay. The biodegradation in poor hay could then be examined in more detail, identifying what organisms are involved in the process and how they are related to the various allergies and infections which can occur in humans and agricultural and domestic animals. Points could then be made which showed that it was generally considered that good hay did not present the same problems as poor hay, however there was some evidence (which could be clearly expanded on and explained) that certain conditions arise in good hay which could be responsible for various pathological disorders. The specific aims of the investigation could then be stated, which involve the study of the types and numbers of microorganisms present in good hay under various storage conditions over a period of 4 months' storage. (Notice that, in the Introduction, little is made of the consequences of the fact that the good hay goes through a similar biodegradative process since it is hoped to develop such points more fully in the Discussion.)

Of course, other approaches could be taken in the Introduction; for example, some researchers favour an approach where they present the aims of the project first and then develop the Introduction from there. Any approach is valid as long as you make absolutely sure that the specific aims for the investigations are clearly presented, justified and that the background to the work is reviewed.

8.7.2   Deciding what points to include in the discussion

The major and central theme of the Discussion should be *the interpretation of the results*. The interpretations presented should be soundly based, thoroughly and carefully explored, well organised in the text of the Discussion, and backed up with evidence from the literature or your own investigations. At all times it is important to avoid being superficial in an explanation, glib, oversimplifying or overstating your case.

Survey your results (using the summarised conclusions) and organise them into groups. It is often useful to identify what are the major results of your project and what are incidental observations or less verified results, which are nevertheless interesting and useful. You might find it appropriate, too, to group your results according to common themes of investigation.

Once the results have been surveyed and organised it should be clear exactly what you will have to interpret, and therefore what interpretations have to be presented in the Discussion. Section 5.7 gives some general advice on interpreting results. Basically, the interpretations must be made in the context of the methods used and in the light of evidence and information from the literature or from investigations within the project. It is therefore generally appropriate to include the following points in your Discussion.

POINTS TO BE INCLUDED IN THE DISCUSSION

A. *AN ASSESSMENT OF THE VALIDITY OF THE METHODS*—The results do depend upon the methods so it is important to make some assessment of the methods in the Discussion, identifying various strengths and weaknesses, such as 'There was a limit placed on the selection of material for assaying because of the restricted availability . . .' Try to be positive in an assessment of the methods and to show clearly how and in what circumstances your work is valid in spite of the shortcomings; for instance, 'Though ideally the animals should have been tested in 100 per cent relative humidity, a high humidity was maintained in the test chamber by always having present some saturated cotton wool. The investigations also have an internal consistency which allows meaningful comparisons to be made between the animals receiving the treatment and the controls . . .' Some ideas of points that could be considered in assessing methods are given in section 8.9.

B. *DISCUSSION OF WHETHER THE RESULTS WERE EXPECTED OR NOT*—It is always useful to indicate whether the results were expected or not, particularly if the investigations were designed to test a specific hypothesis or to verify previous work. It is useful, too, to show clearly how the results match the expectations or do not, for example, 'The rate of nitrite reduction in the light did not exceed the rate of nitrate reduction. This phenomenon is consistent with the activities of the same species reported by Stewart (1978) in which he demonstrated that . . .' or 'The hypothesis that there is a massive input of the olfactory information into the habenula nucleus is not supported by the present work since . . . etc. This would therefore suggest that the habenula does not have the major role in olfactory processing as originally proposed by . . .'

C. *EXPLANATIONS FOR THE RESULTS*—It is necessary to propose detailed explanations of the results obtained. Initially you will have to confirm that the results are not entirely explainable in terms of peculiarities of the method (see section 5.7.2). Having established this, you then need to consider explanations of what is likely to have occurred in the organism to produce the results observed. For example, you might have to explain exactly, and in detail, what could have happened in the microorganisms to prevent them from growing in the presence of antibiotics, or what could have occurred to explain the low level of enzyme activity in a particular tissue, or how a drug could have interacted with a muscle to increase its strength or contraction. By formulating such explanations you will, in essence, be interpreting them and some guidance on this is given in section 5.7.2, which could be referred to. It is generally appropriate to discuss more than one possible explanation for the results and to present arguments for each. For example, 'There are two explanations for the aggregation observed. The aggregation may be influenced by the presence of another species or . . . etc.' Both explanations would then be reviewed. In proposing explanations, particular care must be taken to avoid being glib, superficial or overstating a case—claiming more from the results than can be justified. There is often a temptation in student projects to claim that results are more clear-cut than they really are and that there is only one possible and obvious explanation. Such interpretations are often markedly influenced by what the student believes should have happened (because it makes a neat story), so results are only interpreted in a cursory fashion rather than looked at in detail. Overall an honest, sensible approach is needed in which the limitations are recognised, and this kind of approach will always be to your credit.

D. *PRESENTATION OF THE EVIDENCE TO JUSTIFY THE EXPLANATIONS PROPOSED*—When presenting explanations of results it is always necessary to back them up with evidence which justifies them. It is not enough merely to state an explanation but instead you should show exactly how the explanation fits your results and what justification you have for applying the explanation to them. For example, a statement like 'The results can be explained in terms of the current model for Na^+/K^+ ATPase pump' is not sufficient. It has to be made clear exactly how that explanation fits in with the results observed and what evidence there is for proposing such an explanation. In attempting to justify explanations it is often necessary to give evidence which both supports and which does not support (if applicable) the explanation. In other words, the controversies and arguments for and against a particular stance should be debated. By doing this you will be taking a critical approach in interpreting results. The information you present in reviewing the evidence may come from other workers' results or from additional results in your own project; for instance 'Results presented here indicate that the environment may have induced the trend in stomatal frequency observed. Further evidence for this comes from studies by Brown (1981) in which . . . and the significance of such a development has been examined by Waters (1983) who states that . . . etc.' As you develop the explanations and provide evidence for them, you will need to show where the information comes from by referring to the literature. The next part of the chapter (section 8.8) shows the way this can be done and the examples above also show this.

Having made sure that your Discussion includes these points, which are directly related to your results, it is then useful to consider including the following points which discuss the results obtained from a more general aspect and in the wider context.

WAYS OF BROADENING THE DISCUSSION

A. *SHOW HOW YOUR RESULTS HAVE CONTRIBUTED TO THE BIOLOGICAL FIELD OF STUDY*—You need to place your results in the context of the general area of biology your project is concerned with and to show how your results have contributed to the field or shed light on particular problems within the area of study. Doing this also helps to tie up the Introduction and Discussion. For example, 'The project has shown that the localisation of the enzyme may not be as simple as previously suggested (Ritenour, Joy, Bunning and Hageman, 1977) and that there may be a species of the enzyme associated with the chloroplasts.'

B. *MAKE SUGGESTIONS OF WHAT FURTHER WORK COULD BE CARRIED OUT*—What you are basically doing in explaining your results is formulating new or expanded hypotheses. There will be further experiments which could be carried out to test whether the explanations or hypotheses are correct or reasonable, and in your discussion you should therefore consider these. For example, 'Further studies testing the effects of agonists and antagonists of acetylcholine could provide additional evidence about whether acetylcholine is the likely transmitter at this synapse.' Suggestions for further investigations which could overcome weaknesses in the present study in your project could also be made, such as 'It was not possible to state with certainty that the animals showed a positive choice for the food they had had previous experience of, however a further experiment which could clarify this would be to . . .'

When you have decided what to include in the Discussion organise the points for a natural flow and development. It is best to start from specific points (as they relate to your own results) and then to develop more general themes. It often makes sense to discuss first the Methods (since the results depend directly on these and further discussion can then be made with these in mind). A useful style is to then comment on whether the results were expected or not, followed by various explanations of the results. For each explanation examine it in detail and justify it. Do this by reviewing the evidence, thus identifying strengths and weaknesses. It is often useful to discuss the results either in the same order as the Results are presented in the report or by discussing the more 'important/useful' results first.

Example

The following example briefly outlines the approach taken in the Discussion for a particular student project.

Project title: The effects of metaldehyde on the activity of the foregut of *Arion ater*

Major result: Desynchronisation of regular gut contractions, increased tone of the gut generally and larger contractions, and long lasting effect

Additional results: Different types of cyclical activity were observed; gut activity depended on speed of dissection of preparation and general condition of the tissue; metaldehyde effect appeared as dose dependent and wore off after washing gut; some preliminary results obtained on the effects of postulated transmitters 5HT and Dopamine.

The Discussion concentrates on presenting interpretations of the observed effect of metaldehyde on the gut. Explanations of other results are tied in to this as far as possible. The points developed in the Discussion were as follows:

1. Method examined—the point was made that a commercial preparation of metaldehyde was used rather than a pure solution, so the effects observed may not be entirely attributable to metaldehyde. However, it was pointed out that the project was concerned with examining the toxicity of a molluscicide to determine its effect, so a commercial preparation known to be toxic to the slugs could be used. The point was made that it might be worthwhile testing the effect of metaldehyde in its pure form and comparisons made.

2. Evidence presented and literature cited to show that the results were consistent with the results of other studies on the toxicity, and that they were also consistent with a hypothesis which suggested that the metaldehyde inhibited the gut. Arguments were pursued to show how the results fitted in with this hypothesis.

3. Explanations were proposed to try to explain exactly how the metaldehyde could have had its effect. This meant ideally finding information about the basis of spontaneous activity in the slug gut. However there was little information, so theories relating to spontaneous contractions in other guts and smooth muscles were examined to see if a hypothesis about the action of metaldehyde could be proposed. Evidence was presented to support the explanations from results in the literature and from some of the additional results of the project.

4. Statements about how far the work had gone to solve the original problem were made with strengths and weaknesses shown, and suggestions were given about what further work could be carried out as a development from the results obtained to expand on the work and to combat the weaknesses.

It is most important to take a positive, sensible approach in the Discussion and to show how your results have contributed to the field. Avoid conveying the view that your work is 'useless', 'all wrong', 'that someone did it better than you anyway' or 'that you would just like to forget it all and put it all behind you'. Even when a method was 'wrong' or the results contradict another researcher, careful examination of the investigations can still reveal something useful. It is also important to avoïd just making the Discussion a re-iteration of your results (which is often what students do) or even worse, actually ignoring your results altogether and giving instead a general account of the project area. (If the results are ignored, one is inclined to wonder what all the practical work was carried out for, and this indicates to the examiner that the student either has not understood them or examined them in detail.)

8.8   Showing you know the literature in the project report

First you have to know the literature and chapter 7 gives guides on using the literature. Once you are familiar with the literature you have to show by the style of writing you employ that you *do* know it. This is relatively easy. Each time you make a point quote the reference which either illustrates the point or is the source of the information by giving the author's name or a number. The references quoted should then be listed in the bibliography in alphabetical or numerical order. If there is more than one source, quote each one. Whenever you expand on a point, by for example giving evidence for an explanation of the results, you must give the source of the evidence. When you are making an important point in your Introduction or Discussion, for which there is reference to original material, expand out from the source and give more details. This will indicate that you know what the paper says and you have read it fairly carefully. For example, you might wish to make a statement about the reasons why you are doing the particular investigations in your project. This might have been because of ideas suggested from another investigation; if that is the case give the information from that source in detail. Make sure in quoting references that you have read them and that you include them in the bibliography.

Example

'The acetylcholine esterase activity was also found in the interpeduncular nucleus of the cat and rat (Lewis and Shute, 1967; Krnjevic and Silver, 1966; Kataoka, Nakamura and Hassier, 1973) and was found to be located both intracellularly and extracellularly (Lewis, Shute and Silver, 1967). Lewis and Shute (1967) have shown that the enzyme activity was concentrated around . . . and that . . .', etc.
(that is, details would then be given)

8.9   Demonstrating a critical awareness to your own and other researchers' work in the project report

It is important to show, in the style of writing you employ, that you are able to make a critical assessment of both your own and others' work. This can be difficult as it does depend, to a large extent, on your intellectual ability, your knowledge and experience in general terms and particularly in relation to your own project. However, there are a number of things you can do to try to ensure that you do make some critical appraisal of research considered in the report. You need first to develop a critical approach as far as you can. This basically means routinely assessing the strength and weaknesses of the various aspects of any scientific investigation. Include the original reasons and background of the work, the methods used for collecting and analysing results, and the validity of their interpretation. Throughout this book you have been encouraged to take a critical approach and this can be applied in assessing not only your own work but the work of others. It will also be worthwhile taking note of the following points which provide some additional guidance when attempting critical assessments.

WAYS OF DEMONSTRATING A CRITICAL APPROACH

A. *REFER TO THE LITERATURE*—The papers will show you by example the style for critical appraisal and point out what kinds of criticisms can be made about certain types of work. The literature may also identify certain criticisms which have become fashionable and others which can be made of past work based on current advances in knowledge and understanding of a system or techniques for your project. For example, it has become fashionable to criticise the lack of statistical techniques; it is known now that certain anaesthetics depress or alter nerve cell activity so recordings made from animals with that particular anaesthetic have to be interpreted in the light of current knowledge of the anaesthetic when previous interpretations could not take account of such knowledge. The literature may also have carried out some considerable critical appraisal of previous work for you (up-to-date review articles are particularly useful for this) which you can use in your report.

B. *JUSTIFY ANY POINTS OF VIEW YOU TAKE*—If you are taking a particular point of view, justify it by reviewing the evidence which both supports and does not support that view, making sure the balance favours your view point. For example, if you state that the current investigation will be useful, show why it will; if you state that results could be explained in a certain way, show how.

C. *DO NOT BE 'HIGH HANDED' OR OVERSTATE YOUR CASE*—Do not automatically assume that there is only one correct answer. Recognise that it is more likely that a number of points of view are justified and try to be honest therefore in your appraisals. Do not assume that you have got the answer to everything or nothing—try to be positive and yet show some humility.

D. *REVIEW METHODS AS THEY ARE OFTEN OBVIOUS AREAS OPEN TO CRITICISM*—Basically, you should apply the advice given in chapter 3 on designing investigations to see if a method (yours or another) was justified and suitable for a particular piece of work. The following example is a condensed list of some examples of the kinds of questions you could ask yourself about an investigation to assess the validity of the methods.

Examples

Consider all aspects of the methods from the preparation of the material, the way material was handled, the measurement techniques, the conditions while data were collected, the sampling procedure, to the techniques used for the analysis of the results. You now need to question whether the methods employed in each of these areas were appropriate. For instance, was the sample a random sample, should it be, was it large enough, what preparation procedures were used, how might they have affected the organism, what precautions were taken, why, were they good enough to overcome the problems they were supposed to, what controls were made, were they adequate to determine any effect, do the levels of the experimental factors have any relevance to the organism, does it matter if the experimental factors were very different from what an organism is normally exposed to, were precautions taken to reduce the effects of additional factors, were they good enough, what levels of accuracy were used, were they justified, were they good enough to show up any effect, is the timing of the investigation (daily, yearly) likely to affect the results, could any of the preparation or measurement procedures interfere with the organism's response, what measurement scales were used, was the analysis of the results appropriate for: the sample used; the measurement scale employed; and to test what was wanted (for example, multiple 't' tests are often wrongly used when an analysis of variance should be used instead)?

E. *EXAMINE CONCLUSIONS AND INTERPRETATIONS (EXPLANATIONS) AS THEY ARE OFTEN OPEN TO CRITICISM*—Here you will be looking to see if the interpretations of results are justified (your own and others'). The advice given in section 7.2 on interpretation of results can be usefully applied, and the following condensed list indicates broad areas which can be considered in a critical assessment.

Examples

In making an assessment the first thing to do is to check that the methods are sound since the results depend on these. Then look carefully at the results and confirm for yourself that they really do show what is claimed, for example, a significant rise in phosphate level, a linear relationship, a peak activity, a diurnal rhythm, an inhibition. Do not be too quick to accept, look carefully. In addition, check to see if you can glean any further information from the results. Now look carefully at the explanations of the results. In particular examine the evidence. Make sure that any author referred to has not been misquoted and then, from the literature you have been reading, make two sets of notes — one set to show any evidence which confirms the information and one which does not. Examine the lists and check whether on balance the explanation is reasonable, unreasonable or uncertain. Check to see whether the explanation has been explored in depth. If not, do so and make sure it is still applicable. Finally examine the paper or report to check for inconsistencies in argument.

Having questioned your own and others' work critically, make sure that you include relevant criticisms in the text of the report. This is most easily and neatly done in the Introduction and Discussion where ideas are developed and analysed in some depth. Probably the best way to do it is to inject criticisms in the writings as the points and ideas (which were chosen to be included in each of these sections) are developed. So, for example, if an explanation or justification is being proposed, its strengths and weaknesses could be highlighted and emphasised.

Example

'(explanation→) The glycogen stores may be important as an
energy source (glucose) in times of high metabolic demand or
decreased glucose supply. (support for idea→) Well-fed
specimens do show high glycogen levels which show a decrease
with increased activity and therefore metabolic demand of the
cells. (reservation→) However, the mechanisms relating to
energy supply are not established in invertebrates though
(support→) glucose is known to be the normal substrate in
mammals . . .'

As you develop the various points, make sure that the source of any
information is given by quoting a reference. Overall the following
example shows the kind of writing style which demonstrates a critical
approach and which generally takes account of the kinds of points
made here. Further examples can, of course, be obtained from the
literature.

Example

The effect of extracellular K^+ concentrations was measured on
the uptake of radioactive labelled K^+. (explanation→) The
results of the studies suggested that two separate uptake
mechanisms were involved: one operating at the near
'physiological' level of K^+, and the other at levels much higher
than the normal physiological levels, that is, 20–40 mM up to
80 mM. (criticism→) However, the experiments were carried
out on cultured glial cells whose activity may differ from that
carried out by glial cells *in situ*. (support for criticism→) In
addition, it has been shown previously that the cultured glial
cells are extremely permeable to K^+. (criticism of application
of explanation in the 'real' situation→) Furthermore, the
maximum rate of transport observed in the glial cells may not
be realised *in vivo* since uptake by the cells would cause a drop
in the extracellular K^+ concentration which would not occur
in the experimental situation where an excess of K^+ was
maintained. (general assessment in light of criticism→)
Overall, the results indicate that the glial cells appear to have
the potential to take up excess K^+ that might leak out of the
neurons, but the data do not prove that it really happens.
(criticism of application of explanation generally→) In
addition, other experiments [12, 15] weaken the argument that

> glial cells provide the major mechanism for reducing an
> excessive and maintained elevation of K^+ levels. In these
> studies it was shown that . . ., etc.
> (*Note*: The numbers refer to the references of the work which
> would be shown in the bibliography.)

8.10 Writing up the remaining sections

8.10.1 Bibliography

The bibliography should consist of a list of all the references you have
referred to in the text. The list can be alphabetically or numerically
arranged, and it is probably easier to do the former. The text then
refers to authors and the source can be obtained from the bibliography. Some examples of how a reference is quoted are given below.

(i) *A paper from a journal*—Author, date in brackets, title, journal,
   volume underlined or in bold type, inclusive pages.
   For example, TAYLOR T. N. (1982) The origin of land plants: A
   paleolobiological perspective. Taxon **31**, 155–177.

(ii) *A book*—Author, date in brackets, title of book, publisher, place
   of publication.
   For example WOODFORD, F. P. ed. (1968) Scientific writing
   for graduate students. Macmillan: London.

(iii) *An article or chapter in a book*—Author of article, date in
   brackets, title of article, book title, source of article in the book,
   publisher of book and place of publication.
   For example, LUMSDEN, R. D. and SPECIAN, R. D. (1980)
   The morphology, histology and fine structure of the adult stage
   of the cyclophyllidean tapeworm, *Hymenolepis diminuta*. In
   Biology of the tapeworm *Hymenolepis diminuta*, ed. Arai, H. P.
   pp. 157–280. Academic Press: New York.

If one author listed in the bibliography has a number of papers,
these are usually listed in chronological order with the most recent
occurring first. If there are a number for each year then they are
identified by for example 1975a, 1975b, 1975c.

8.10.2 Summary

The summary should be a concise account of what you did, what
results were obtained and briefly how they were discussed. Summaries

of research papers are often given in the form of an abstract of the paper. The summary is usefully placed at the beginning of the report, after the title page, as a guide for your reader.

8.10.3 Acknowledgements

It is customary to acknowledge the help given by supervisors, friends, technicians, and typists, the department in which the work was carried out and the grant authority supporting the work if relevant. The acknowledgements are often placed at the end of a report.

Index